Artist Management for the Music Business

Volunteering for NERFA is open until
nerfa@comcast.net - interested in
volunteering/people need roomates?

Caffe Lena - Saratoga   (Open mics)
Stanford NY   Arts
Minekill   Hudson River
Fultonville

PRO's are necessary - registration for any
airplay royalties.

All Songs in Altamont - allsongs.org June 27th
Clearwaterfestival.org - June 20 + 21st 28th
       Croton + hudsons
Philly Folk Fest - volunteering? - August 15th

House Concerts - libraries? meeting room of
ambulance corps. Suggested donation/$10
friends in another town - build up an email
list Middleburgh Telephone Company.
Concert Thursday Horowitz + Malkine.

Benefit Concerts - Helping Hands Concert -   Hillsdale
falconridgefolk.com Falcon Ridge Folk Fest
                   - July 31st / August 1st + 2nd

- 50/50 at Gate.
- Series or Major Benefit
  Schoharie County Arts Grant
- Renee Neid - apply to
- Schoharie countyartsgrant @ gmail.com
  Peter Yarrow - email management.

- Humour songs - audiences go out to shows to be entertained
- Audience Interaction is important
- Do at least 1 or 2 cover songs
  - show off your vocal range.
- Showcase - do your most reactive songs

Open mics - test out your new songs
          audience reaction.

Rick Brodsky - spoke to Sonny / would like to be on crew.
Keep all receipts for business expenses
lunches / gas / food
look into write offs

Artist Management for the Music Business

Paul Allen

Focal Press
Taylor & Francis Group

NEW YORK AND LONDON

First published 2007
This edition published 2013 by Focal Press
70 Blanchard Road, Suite 402, Burlington, MA 01803

Simultaneously published in the UK
by Focal Press
2 Park Square, Milton Park, Abingdon, Oxon OX14 4RN

Focal Press is an imprint of the Taylor & Francis Group, an informa business

Notices
Knowledge and best practice in this field are constantly changing. As new research and experience broaden our understanding, changes in research methods, professional practices, or medical treatment may become necessary.

Practitioners and researchers must always rely on their own experience and knowledge in evaluating and using any information, methods, compounds, or experiments described herein. In using such information or methods they should be mindful of their own safety and the safety of others, including parties for whom they have a professional responsibility.

Product or corporate names may be trademarks or registered trademarks, and are used only for identification and explanation without intent to infringe.

Library of Congress Cataloging-in-Publication Data
Application submitted

ISBN-13: 978-0-240-81501-5 (pbk)
ISBN-13: 978-0-240-81502-2 (ebk)

Contents

Acknowledgments

I extend my personal and deepest thanks to industry professionals and colleagues who helped guide my work on this book. This group of special people includes Cosette Collier, Mike Milom, John Beiter, Tom Baldrica, David Ross, Hal M. Newman, Amy Macy, Chris Palmer, Paul Fischer, Chad Campbell, Trudy Lartz, Jeff Walker, Troy Festervand, Geoff Hull, Tom Hutchison, Richard Barnet, Jon Romero, Jeff Leeds, Lee Logan, Bill Mayne, Larry Pareigis, Mike Dungan, Charlie Monk, Joni Foraker, Tandy Rice, Denise Nichols, Clarence Spalding, Jim Beavers, Dan Franz, Nathan Brenner, Catharine Steers, David Corlew, Mike Alleyne, Melissa Wald, Shelia Biddy, John Dougan, David Bowers, Matthew O'Brien, and career managers in the music business who have been so generous with their time and insight – plus the countless others I have encountered during my career who have allowed me to learn by being involved in their careers, especially Cindy.

Introduction

This book is intended to be the definitive guide to the student of management of artists in the music business, as well as to those seeking to become professional artist managers. Some of the tools developed for this book are found nowhere else, and active artist managers will find them to be helpful planning and organization tools. The companion website for this book, *http://www.artistmanagementonline. com*, is a continuing resource for both the artist manager and artists, and includes a directory of artist management firms, advice, and links to help the manager be a more effective manager in the music business.

Information from this book has been drawn from the experiences of many who work or have worked as artist managers, as well as from the author's career managing people, assets, companies, organizations, projects, performers, and performances. My wish is that the words that follow will be your guide; my dream is that they will be an inspiration.

Professional artist management and its principles

CONSIDERING ARTIST MANAGEMENT AS A PROFESSION

Whatever your title—manager, personal manager, artist manager, brand manager, or representative—managing artists in the music business means that you are becoming a part of every facet of someone else's life. There is virtually no aspect of the professional and personal corners of an artist's life that a manager doesn't encounter on a regular basis. Helping direct the career success of an artist requires significant involvement in their life. A manager who is new to the profession will find it to be immensely time-consuming and slow to deliver rewards, yet energizing with its fast pace and regular challenges.

The music business swirls in its own continuous change, and the result has been the shrinking influence of large record labels in the careers of artists. Instead of being an adjunct to the work of the label on behalf of the artist, managers now find themselves at the hub of the artist's career, providing many of the services formerly handled by labels, and they wield considerable control over the success of music careers. Labels are now selling less music, and in response, continuing to cut overhead in an attempt to remain in business. Labels have traditionally cornered the distribution of music, but online sales have significantly reduced the need for a distribution system designed to deliver pallets of boxed physical product to warehouses. As traditional labels see their roles diminishing, the opportunities for managers of recording artists have never had a brighter future—today, many management companies provide most of the services of a record label and may eventually replace the label by recording and marketing music on behalf of their artists. Clearly, a career in artist management requires a continuing—almost daily—education by paying attention to what is happening in the music business and other industries and events that affect it, and what that means to the artists they manage.

Artists in the music business are sometimes managed by attorneys. But there is a reason why professional managers are the best choice for artists to manage their careers. One of the top entertainment attorneys in the country once told over lunch that attorneys are not necessarily the best choices to provide career management to artists, primarily because of their conservative nature as practicing professionals. Lawyers are disposed to advise their clients on ways to conduct business without creating conflict. Today's artist manager, in order to stand as the strongest advocate possible for an artist in a highly competitive industry, must be able to push that advocacy to the limit—without overstepping the boundary of business ethics. Attorneys have become effective managers, but doing so generally requires that they step away from practicing law.

Be creative, informed, and connected

Today's artist managers must be willing to encourage their artists to take calculated risks and to support their clients when they do. This doesn't mean they take chances with an artist's career. Rather, they involve the artist in promotional ideas that get the artists outside of their comfort zone and open up opportunities to help them reach their goals.

While you're managing someone else's career, you also must manage your own. That means you must keep up to date on the entire music business. Certainly technology and changes in the legal environment of the music business create new directions and challenges for artists, but they also provide opportunities. As a manager, you must be aware of trends and how they affect artists on your management roster. That means regularly reading publications like *Billboard* and *Pollstar*, attending industry conventions, and subscribing to online industry headline services like those provided by *Billboard.biz* and *AllAccess.com*.

Veteran artist manager Ken Kragen titled a book he cowrote *Life Is a Contact Sport*; in it, he discusses the importance of developing and servicing a personal network of contacts. Being able to get that telephone call returned is among the most important assets an artist manager has. Without the connections—either direct or indirect—it is difficult to get business done on behalf of the artist. For the aspiring artist manager, yesterday wasn't too soon to begin building that network.

Understand people and business

Developing meaningful interpersonal relationships can be challenging, but it is more important to the artist manager than any other skill or talent. The work of an artist in the music business is a web of negotiated deals that requires the manager to have patience, an understanding of human nature, great communication skills, and a solid reputation of dependability. Each of these traits requires cultivation, but each will also become the foundation of a successful career in management.

Aside from the music, business is the other constant in the career of an artist manager. As surely as music connects with an individual's passion, it doesn't become commercial until it's good for business. And to conduct business on behalf of the artist, the artist manager must develop an understanding of team building, marketing, budgeting, and sales as they apply to the income streams available to the artist.

There are frequent references in this book to a *360 deal*, also known as a *multiple rights recording contract*. This term means that a company/label is entitled to a percentage of some or all of the income streams of an artist in the music business. It is most often applied to recording contracts that give labels part of the nontraditional earnings of new artists that they sign, such as part of their merchandise or ticket sales in addition to profits from marketing their recorded music. As you read this book, you will see that the artist manager in today's music business is actually in the best position to direct, profit from, and control 360° of the artist's career.

Let's begin with a look at the business and science of management. For the reader who is relatively new to management science, this chapter is the starting point. It puts the balance of this artist management book into the context of the basic principles of management. As you will see in the chapters that follow, managing an artist in the music industry uses science, business, and a good measure of creativity to achieve success for clients. This is especially true of the music business because of its nature as an industry that offers high rewards for the relatively few who become successful—where success is often measured by affluence.

As we consider principles of management, it is important to understand that the goals of artist management are different from those of other areas of the music business. Record companies are in the business of marketing and selling recorded music and related products. Traditional radio companies are in the business of building audiences to lease to advertisers, who in turn purchase spot advertising from the radio station that airs within the programming. Concert promoters present a live entertainment experience. Artist managers are in the business of developing long-term careers for their artists, which includes touring, merchandising, sponsorships, licensing, recording, songwriting, and the full exploitation of all of their talents. Some artist management companies combine all of these functions for the artist under one umbrella.

As we look at management principles, it is important to understand that the work of an artist manager in the music business is somewhat different from other kinds of managers. The traditional relationship a manager has with an employer is one that has a reporting hierarchy, and by definition is very structured and "corporate" in nature. For example, a copy writer reports to a creative services manager, who reports to the director of marketing, who reports to the VP of sales, who reports to the president; an artist manager reports to only the artist. Traditional managers use resources of owners to ultimately sell goods or provide services for a profit, and in many ways, that is what the artist manager does.

Much of the work of an artist manager is product development, sales and promotion, planning, and managing the work of the team around the artist. The relationship between the artist (employer) and the manager (employee) is considerably closer than that of typical managers in business, and is much more like a partnership. The level of trust and the strength of the relationship between the two are often compared to those found in successful marriages. That kind of association of a manager with an employer is rarely found in the business world. However, there are times when an artist manager takes on most of the traditional roles of management as he or she oversees the management of the artist as a brand, with the artist being a creator of art and entertainment experiences.

FUNCTIONS OF MANAGEMENT

Nearly every text, research paper, and discussion on the topic of management embraces four classic functions: planning, organizing, directing, and controlling. These functions all apply to the work of the artist manager in the music business.

Planning

The difference between success and failure in any endeavor can often be tied to planning. Luck by itself can sometimes deliver success, but coupling it with a well-designed plan can put the manager in a position to take advantage of opportunities when they present themselves. It is very satisfying when opportunity opens a door to implement an active plan to take advantage of it. For example, young Josh Groban was asked by award-winning producer David Foster to replace an ailing Andrea Bocelli in the 1999 Grammy television rehearsals with Celine Dion. His performance at the rehearsal was powerful enough to help launch his multiplatinum recording career as an artist. Groban's planning and preparation for a career as an artist put him in a position to benefit from the lucky timing of Foster's telephone call. Eight years later, he had the top-selling album in the world, and by 2011 the career sales of his recorded music approached a quarter-billion dollars.

When a leader or manager identifies worthy goals, he or she often collaborates with stakeholders to develop a set of logical steps to achieve them. Those steps, or plans, become the framework for successfully meeting goals. Dr. Carter McNamara puts it very well when he says, "Planning is identifying where you want to go, why you want to go there, how you will get there, what you need in order to get there, and how you will know if you're there or not" (2006). It is easy to see why planning is often viewed as a road map that helps define the route to success. A career plan results from collaboration between the manager and the artist, which provides direction and milestones to reach goals. This book frequently addresses career planning essentials.

Organizing

Organizing the manager's work is closely tied to the planning function. Organizing is assembling the necessary resources to carry out a plan and to put those resources into a logical order. It also involves defining the responsibilities of the artist's team, and managing everyone's time for efficiency—especially the artist's. The manager allocates the amount of time necessary to follow each step of a plan to get the intended results.

The manager of any enterprise also seeks funding or financing necessary to pay for the plan. The grandest example—long before the great recession—is Chrysler Chairman Lee Iacocca's successful pursuit of hundreds of millions of dollars in loans from the U.S. government in the late 1970s to save his ailing company from bankruptcy. For the artist manager, financing and funding the plan for a new artist's career could include a combination of an accelerated touring schedule, finding sponsors, relying on assistance from fans who offer financial help, asking friends and family, and helping the artist secure loans. Managers also recruit and employ labor and expertise to put the plan into operation and to see it through to its success.

The manager of an artist in the music business forecasts the need for members of the artist's team, and plans for the time when their services will become an expense to the operating budget for the artist. The manager also draws any other necessary resources together, creates a logical structure for the organization of those resources, develops a career plan, and executes it. An artist looks to the manager to take the chaos of a prospective career and organize it into the prospect for success.

Leading and directing

Managers provide a leadership function for the artist and their team by ensuring that the talents and energy of the team are directed toward the career success of the artist. The work of a manager in directing activities is to take the resources needed to reach goals and use them efficiently to achieve success. For example, an artist manager often hires a company or an employee to oversee the successful application of new media strategies to the promotional plan for the artist's career. (Promotion and marketing are elements of the overall career plan.) This means the manager coordinates the energies of the professional team members working toward the artist's career goals, monitors income and the expenditure of funds, and plans and manages time. And all of this work requires that the artist manager keeps everyone directed toward achieving the career plan objectives.

An artist manager sets up a team of support for the artist. Some are on the active payroll and others are used to support the plan on an as-needed basis. The term "team" as applied to the group of professionals who support the artist is indeed a group with a common goal: a successful career for the artist. However, it is rare that the team as a group will assemble for a meeting about the artist. Instead, the artist manager provides each with continuous communication about the activities of the artist, and draws expertise or assistance from each member of the artist's support team as their help is needed. And members of the team communicate with each other as necessary. For example, the artist's booking agent forwards budgets and offers from promoters to the artist's manager, business manager, and perhaps accountant, and then waits for feedback from each.

Controlling

Any manager who has created a plan follows its implementation by controlling all of the resources required to achieve the goals of the plan. When the resources (time, people, equipment, financing) have been assembled and the plan is under-way, the manager monitors how effectively the plan is being carried out and makes any necessary adjustments in order to be efficient with the use of resources and to be effective in advancing the plan.

The business of managing an artist in the competitive world of the music business means developing strategic plans in an effort to control as much of the artist's devel-oping career as possible. The manager must be realistic in what he or she feels able to control, but it also means that he or she must be flexible enough in encounters with reality to adjust to the circumstances. For example, a manager should anticipate that a new and promising artist will not be able to give a powerful performance at each audition, and should be prepared to put the most positive "spin" on the result.

ARTIST MANAGEMENT SKILLS AND PERSONAL TRAITS

The management skills discussed in this section are an indication of the breadth of the practical understanding of people and the music business required by an artist man-ager. For the prospective artist manager, these skills constitute a guide to learning; for the active manager, they are an affirmation of the truly special talents of manage-ment professionals who guide the careers of artists in the music business.

Understanding human nature

Managing an artist's career requires interaction with people of all personality types and under many pleasant—and some unpleasant—circumstances. Studying why people react as they do to events in their lives is the best way for the prospective artist manager to learn. Among the most challenging can be those times when it is necessary to manage around the ego of another. The music business is one in which egos thrive, with many trying to be "somebody" or trying to assert that they already are somebody, and often these very people can be the gatekeepers to the next step in the artist's career. Carefully playing into the ego in this circumstance is an effective way of using human nature to the manager's advantage. Before playing into a gatekeeper's ego though, the manager will need to know enough about the individual and his or her ego drivers. That is, knowing the particular human will let you know his or her human nature. For example, if the manager is trying to recruit the services of a top publicist who is reluctant to take on an additional client, that conversation must include references to a specific artist promotional campaign in which the publicist was a key component in another artist's success. Explain that you want nothing less for your client. The manager should also acknowledge any awards or special recognition the publicist has received resulting from their recent work. An understanding that personal achievement is important to professionals gives the manager an opportunity to service the gatekeeper's ego and open a conversation.

Leadership

Leadership is an important skill, or trait, of an artist manager at the beginning of an artist's career or at the beginning of the relationship. The influence of an artist manager on the early planning and development of an artist's career is what helps the artist to develop a focus and an organized purpose. At this point, we are merely acknowledging the importance of leadership skills to help direct an artist's career, but we take a deeper look into both leadership and coaching and how they apply to a career in management in Chapter 13.

Coaching

Coaching skills for the artist manager are closely related to those of leadership. Leadership seeks to guide the broader, long-term goals of the artist's career, but coaching involves short-term work toward an outcome that improves the artistry of the artist. For example, the manager-coach helps the artist improve a competency such as being able to develop more animation in his or her stage presence. Acquiring the skills of a coach requires that the artist manager study others who coach (regardless of the sport or profession), and draw from observed styles and techniques. Certainly, most managers will be unable to coach an artist in all of the creative and technical areas necessary, so it is important that the manager has a good network of specialists who might include vocal coaches, physical trainers, stage direction coaches, interview coaches, and more.

Networking

If an artist manager in the music business cannot get a call returned, he or she is ineffective, so building a network of contacts and relationships early in a career is extremely important. To build a network, a manager must be willing to become involved in both the business and social sides of the music industry. A starting point for a new manager is to become a junior associate with an existing firm where a personal network of contacts can be developed. Drawing from the networking resources of an established management company is a convenient way to become recognized as a manager and build a reputation.

Industry events—such as conventions, awards shows, conferences, and seminars—are good places to meet key players who may be helpful in the careers of both the manager and their artists. Examples would include the annual *College Music Journal* (CMJ) conference for U.S. college radio stations, the *Billboard* Hip Hop conference in Atlanta, the *Pollstar* convention, the Country Radio Seminar held each year in Nashville, and events presented by the UK's MusicTank. When participating in events like these, it is important to set personal objectives to optimize the time and money spent invested to attend, because there are numerous social and showcase distractions that can take important time away from intended business purposes. For example, set a goal of becoming introduced to ten key people, and then set up a luncheon meeting with at least two of them to get to know them better and to draw them closer into a personal network of professional contacts. Be creative and aggressive in building a business network because the relationships you develop are key to your personal career success and to the success of your clients.

Social

Closely related to networking are social skills. For artist managers, being social means having a congenial and approachable style that gives the appearance of being comfortable in many quasisocial and business settings. Appearing comfortable in a social setting—whether the manager is or isn't actually comfortable—gives an outward look of confidence. Managers should know how to engage others in conversations appropriate for the situation, how to begin and end those conversations, and how to make them productive. For example, a time-tested conversation opener is, "How are things with you?" which lets you know what is important at the moment to the other people and lets them talk about one of their favorite subjects—themselves. Another social skill is knowing the appropriate attire for different types of business meetings and events. Not every artist manager enjoys social settings, but being a part of them and looking the part are requirements of the job.

Being aware of political circumstances among companies and personalities within the music business is important for the artist manager to keep in mind. In this circumstance, "political" does not refer to a party or candidate affiliation; it has to do with the negative and positive business relationships between people in the music industry, and how they affect the ability of the artist manager to conduct business on behalf of his or her clients. Understanding business alliances between individuals and companies can help the manager save time by avoiding unproductive pursuits and focusing on those with a likely positive outcome. For example, the manager may not want to hire a publicist for one of his or her clients who was fired for cause from a prospective record company. It might make good business sense, but the politics of the situation could make the development of a record deal with that company difficult.

Communication

Having good communication skills means the manager knows how to continuously connect with others, choosing between the written letter, email, tweets, instant and text messaging, telephone, fax, social networks, and every wireless device that will ever be invented. The manager will be interacting with all parts of the music business and must be prepared to use the communication tools that are favored by that sector of the industry. For example, many radio promotion people who work for record labels depend on wireless devices to continuously communicate with the label and radio stations. Managers must also be advisors to their artists on which forms of communication they should use, and when and with whom they should use them.

Each form of communication has its own protocol when used in the business setting, and the artist manager should be sensitive about when and how to use each. Former Sony music executive Jack Lameier, for example, championed a voice mail courtesy that urges callers to office phones to leave their telephone number twice to prevent having to play a long message more than once to retrieve the number. Wireless phones display telephone numbers, but many office systems do not. If you don't know which kind of device you are calling, leave your name and phone number twice. Email is another communication tool that sometimes takes on the loosely written style of a text message, but it is important to understand that email has become a semiformal medium of business communication that doesn't require smiley faces. As author of this book, I receive frequent emails from aspiring artists, and it is easy to tell from their email style which ones are prepared for the business of music and which are not.

Other skills

Artist managers spend much of their time planning and organizing on behalf of their artists. Later chapters in this book describe in great detail the ways these skills are applied to the manager's work.

The best managers also work creatively. The hugely competitive nature of the music business requires that the managers must push their own creative skills to their limits in order to advocate on behalf of their clients. Simply doing what every other manager does is not enough to gain recognition for an artist's talents and potential. An example of using creativity on behalf of artists is a bold idea by Big Machine Records CEO Scott Borchetta. Early in his career, his idea to attract attention for one of the artists he was promoting at an industry convention was to hire an entire high school band to march into Nashville's Wild Horse Saloon.

He and his artist received considerable attention and left much of his competition saying, "Why didn't I think of that?" Fast forward to today: his creative skills now guide the career of one of the biggest stars on the planet, Taylor Swift.

Artist managers are persistent. Persistence in this environment is a quiet determination without being pushy. Remember that to be effective, the manager must navigate around gatekeepers, and an overly aggressive style can be offensive to some of those whose help is needed on behalf of the manager's artists. However, the manager who quietly waits for an email or for the telephone to ring has clients whose careers will never ignite.

The best managers understand salesmanship and use those skills to create an interest by others in their clients. Specifically, they are the chief advocates, promoters, and cheerleaders in the music business for their artists. Through salesmanship, they use persuasion to influence and motivate industry gatekeepers on behalf of their artists.

Artist managers have a good sense of business, and are good at budgeting both time and money. Effective and efficient use of time can keep goals, strategies, and tactics on track. Planning the financial aspects of an artist's career assures that the necessary funding is available when it is needed to support the career plan.

Artist managers have the necessary technical skills to deal with everyone from new media planning to the sound technician at an artist's performance. It does not mean that they must be able to perform the tasks, but must know enough about the technical aspects of the tools that support an artist's career to ensure they hire the best they can afford.

Managers are also skilled at keeping themselves goal- and results-oriented. This means they keep their focus on the artist's career, and they keep the artist from being distracted by the frustrations that go along with pursuing a competitive, high-rewards career. Artists, too, must maintain a focus on goals—and it is the challenge of the manager to keep them motivated.

Problem solving, the final skill covered, is perhaps the most important. This can include defusing conflicts, resolving interpersonal issues, finding alternative ways to get results, and being the "go to" person when major components of an event for the artist begin to fall apart. When confidence in every other possible solution is shaken, the manager is prepared to take action and solve it. And almost as important as problem solving is the ability to find problems and deal with them *before* they become something to which people must react.

BUILDING A CAREER IN ARTIST MANAGEMENT

An artist manager grows in expertise in many of the ways managers in other industries do, and some approaches for growth are especially applicable to the work of someone in the music business. Here are a few such strategies that will help build a manager's position within the profession:

- Give yourself a good self-assessment by looking at the strengths you have and how you will use those to your and your clients' best advantage.

- Under every circumstance, be professional. There is no better way to build and strengthen a reputation.
- If you work within an artist management firm, do all you can for the success of the company, not just for your clients.
- Build and maintain a network of contacts. That means calling on your contacts for assistance when you need it, but also offering help when they don't expect it.
- Be sure the management firm you choose is a good match of styles, and be sure the artists you manage are a good fit with your personality and style.
- Look for a mentor who can guide your development and be a sounding board for issues you face in management. The best mentors keep you out of the quicksand.
- The smartest managers recognize that for genuine growth, a career requires continuous education. Attend an occasional seminar or take a course online, read relevant blogs, and bookmark websites that can keep you current in the music business.
- Know the business etiquette for the music industry.
- Be prepared for career setbacks and brace yourself for recovery.
- Keep an ongoing record of the things you achieve for yourself and for your clients, and document your public service work as a reminder to yourself where you contribute time.
- Be a mentor to your eventual replacement, even though you and others think you are irreplaceable.
- Be better than your competing artist managers, and become known as an expert.
- Look successful even when your career is still a work in progress.

Reference

McNamara, C. (2006). *Planning in Organizations*. http://www.managementhelp.org.

Preparing to manage

2

The short course on preparing to manage artists is this: get some education, get some experience, build a network, find some clients, and then go to work managing. This chapter includes details on some of those steps, but it also includes a deeper look into the realities of the business, and the character of the relationships that a manager develops in the industry. And most importantly, it will help managers understand realities that have an impact on their effectiveness at advancing the careers of artist clients.

MANAGEMENT IS PART OF A BIG BUSINESS

Managing artists in the music business is done within the context of the general business environment, which is influenced every day by the economy, technology, politics, social and cultural factors, the price of energy, and international tension.

Likewise, the practice of artist management is done within the environment of the music industry, which deals with issues such as illegal file sharing, shrinking market shares, competition by other entertainment media, and rapidly advancing technology. For example, Chris Anderson's disputed idea of *The Long Tail* (2006) suggests to the music business that the industry is moving from being one that is hit-driven to one that serves countless niches of appetites for unique music. Anderson's theory is fueled by the speed of the development of technology and the speed of its adoption by consumers of music. Alternatively, Anita Elberse says in the August 2008 issue of *Harvard Business Review* her research demonstrates that entertainment continues to be driven by music and motion pictures that are defined as hits that have broad appeal to consumers.

A few years ago Live Nation was researching ways to expand its business and discovered that artist managers and booking agents globally earn a combined $10 billion (Leeds, 2006). Agents and managers are responsible for connecting artists with career income, and both earn a percentage of the economic activity they generate for artists and for themselves.

The International Federation of Phonographic Industry provided its estimates of earnings by the music business from a global perspective (see Table 2.1), showing its broad impact as well as areas of opportunity to managers of artists' careers.

An artist manager typically earns 10–25% of the earnings of an artist in the music business. However, if no artists under contract with the manager are earning

Table 2.1 The International Music Business

Music Business Segment	Annual Sales
Radio Advertising Revenues	$32.5 billion
Recorded Music Retail Sales	$27.8 billion
Audio Home Systems	$25 billion
Portable Digital Players	$24.2 billion
Live Music Sector	$21.6 billion
Musical Instrument Sales	$16.4 billion
Music TV/Magazine Advertising Revenues	$9 billion
Music Related Video Games Sales	$4.5 billion
Publishing	$4.6 billion
Performance Rights Market	$1.7 billion

Source: IFPI, 2010 Investing in Music.

enough to cover their own expenses, the reality of when commissionable money flows as income to the artist can be a shock to new managers if they're not prepared for it. Recording royalties from large labels can be two years or more from the release date (assuming the album recoups advances to the artist), and income from the sale of tickets to performances can be minimal to nonexistent at first. Publishing royalties can be slow to develop if the artist is new to songwriting. So a new manager should be prepared to finance his or her own management business for three to five years. There are other ways to enter the business of being a professional artist manager, which are discussed in this chapter. Meanwhile, a more detailed analysis of a manager's compensation is presented in Chapter 5, where we examine the contract between the artist and the manager.

ACQUIRING THE KNOWLEDGE

The aspiring artist manager needs a basic understanding of the entire music industry. Without it, the manager will never know about opportunities that he or she has missed on behalf of artist clients. Earning a degree in business is very useful—especially one that emphasizes marketing, brand management, and salesmanship. Earning a degree with an emphasis in the recording industry can give a graduate a very broad, contemporary look at the business that many active players in the industry do not have. A degree in law is helpful, as is a masters in business administration, although an undergraduate degree in business gives the manager a good general working vocabulary and understanding of the business world, which help effectively manage artists.

Aside from the classroom, an especially effective way to acquire knowledge is to work for an artist management firm. Opportunities at management companies are limited, so a very basic, entry-level position may be the best that is available, but the experience will permit the aspiring manager a place to develop a hands-on understanding of the artist management business, and help build that important personal network of industry contacts.

Another option is to seek an internship or apprenticeship with a management and production company that is willing to put the aspiring manager on the road for a season of touring. Many of the basic issues that a manager regularly handles come up on a tour, and one can get a good education about management issues by being in that environment.

UNDERSTANDING THE NATURE OF ARTIST MANAGEMENT

Any business is primarily providing a form of customer service to a target market, and artist managers must keep a good working definition of the customers that the artist serves. Certainly, the fans who buy tickets, recordings, merchandise, and digital products are first on the list. However, others that artists and their managers serve include concert promoters, talent bookers, sponsors, members of the artist's team, and anyone else who relies on the artist to provide something. Striving to promote good relationships with people who rely on the artist creates the kind of reputation that draws and keeps the kinds of professionals who are willing to invest their energies in the success of the artist's career.

Making decisions

Decision making about an artist's career certainly has to do with business, but there are also decisions to be made that deal with art and artistry. As tough as some of those decisions are for the artist manager, management icon Lee Iacocca puts it best: "If I had to sum up in a word what makes a good manager, I'd say decisiveness" (Iacocca and Novak, 37). Often being decisive is easier to say than to do. For example, choosing the commercial viability of a single recording from a ten-cut album of music requires a certain intuition based on experience. In the music business, it often is referred to as making a decision based on your gut feeling. Former BMG Music Group CEO Clive Davis was said to have golden ears, because into his seventies, he continued to hear the sound of commercial musical success when choosing which artists to sign. Although Davis used a different part of his anatomy, it is the same concept as using your gut to guide your decisions. Artist management today requires professionals to step away from the standard strategies that have been used in the past to manage careers of artists in the music business. Some things remain as they always have been, but there will be times when a manager must rely on that special inner feeling to direct the artist to an opportunity because it feels right, or, to guide them away from it because it does not feel right. There are no criteria for this kind of guidance. You just need to follow your gut.

Managing pressure

The stress that comes with managing an artist is set in motion when the manager begins promoting the artist's career. When launching a career for someone else, there is constant pressure to plan and to try alternatives that will work. On the other hand, when an artist has become popular, there are demands on his or her time that the manager must try to meet, because an artist's career in the music business has a limited shelf life, and there is an urgency to draw as much from it as possible while demand is there.

Artist management causes pressure for those whose style prefers order and predictability. Although an artist's manager is an advocate and is viewed as someone who makes things happen, managers cannot predict when people will respond. That means that the manager must always be prepared to react, regardless of the time of day or the day of the week. So an artist manager does not have set office hours with weekends and holidays off and a guaranteed two-week vacation each year.

Artist managers are ultimately responsible for the success of every performance an artist makes, and many of those performances are rehearsed during the week and performed over weekends. When you add in the other dimensions of an artist's career and multiply it times the number of artists in a manager's portfolio, it is easy to see that 24/7 easily could become 25/8 if it were possible.

With the stress of the work, it becomes very important and necessary to balance personal life with work as an artist manager. This book advises and guides artist managers in the creation of plans on behalf of artist clients; however, actively planning for personal time can keep artist managers refreshed and have them prepared to handle the business of managing careers. Planning for that balance between work and personal life can minimize the burnout that can result without it. Occasional long weekends are equivalent to a power nap—they're a quick refresher and then it's back to work. Plan for longer breaks. Block out the time and take at least a week away from management responsibilities. Without an extended break from the office, it is impossible to genuinely break the frenetic cadence of the office environment, with its cache of wireless and electronic devices that keep you lashed to your work. Put your watch in a drawer and give your digital assistant to your human assistant for a week.

Perhaps the best strategy an artist manager can use to handle the pressures of managing within the music business is to learn to put stressors in perspective. The most effective artist managers are those who are not prone to take the actions or inactions of others personally. In nearly all circumstances, the response or lack of response by others has little to do with your work as the artist's manager; rather, it has to do with *their* personal and professional agendas.

Knowing to expect some of the stressors that accompany a career of artist management can help a manager approach issues professionally and with the aplomb that keeps them from becoming personal. As you consider the realities outlined in this chapter, remember that they usually are business—not personal—and these are specific realities that can cause the manager stress and consternation *only* if they are permitted to. Following are some of the realities in the music business that have the potential to create stress.

Reality one

A lot of people will say "no," and a lot of gatekeepers will seem immovable. The music business on its grandest scale offers the promise of great financial rewards for the relative few who are able to connect with a large audience. This means that there are countless thousands of talented people seeking ways to access gatekeepers who can give career opportunities to artists in the music business. When they say "no," it is not a personal response to the manager. It simply means that the manager must find a way to get past the gatekeeper, or must pursue another one who is more likely to be a favorable audience. Finding out what the gatekeeper needs and using it as a way to gain access is one strategy (Kragen and Graham, 1994). For example, personal or executive assistants are essential gatekeepers in any company and are often underappreciated for their contributions. They have a need to be recognized for their value to the company, so the manager who takes time for a brief visit with them implies to the assistant their importance and builds an ally who can perhaps open other gates within the company.

Another way to get past gatekeepers is to find someone in your network who can open that door for you. Build the network and use it.

Reality two

You are a "player" in the music business only if you are relevant. Relevance has to do with your current activity in the industry—specifically, in the case of an artist manager, this is defined by the artist management firm for which you work, or by the artists you manage. If your telephone calls are not returned or your emails are not acknowledged, it is not personal. It just means that you haven't developed the perception that you offer something that will improve the other person's business.

I emphasize throughout this book that every business encounter a manager has on behalf of artists they manage isn't about the artists. It's always about the agenda and needs of the other person. You must be able to demonstrate to others that you and your artists are relevant to continued success of the other person's business.

Reality three

Keeping the spirits of your artists up during a continuing career roller coaster will be draining for you. Understand that the pursuit of opportunities for your artists will include many rejections, due primarily to competition within the industry. Even if he or she knows that rejections are a regular part of the music business can still be a defeat for the manager. And then the manager is the one who must pass the difficult news along to the artist. Finding a way to cope with disappointment at the personal level and then being able to find a way to keep up the spirits of the artist is always a challenge for the artist manager.

Reality four

People will string you along. Early in the author's career in the music business a prominent music industry publisher advised, "Don't pet stray dogs." His point was that you can waste a lot of time by giving advice and befriending individuals who are not in the industry yet and have a long way to go before they will have any chance at success. Some in the industry will tell others whatever is necessary to get them off the telephone or off their doorstep if they think the individual has nothing to contribute to their business. A frequent tactic is to be told "They're tied up in a meeting. May I put you into their voice mail?" You leave your message and there is no chance that your call will be retrieved and returned. It is not personal. They just don't know how you can contribute to the success of their business and don't have the time to figure it out. The second time you hear from a personal contact saying, "I'll get back to you in a couple of days on that" is the time to move on. It's a not-so-subtle way of saying, "Don't call me again." Don't allow people to string you along—they're wasting your time. (Author's note: Petting stray dogs with candid and genuine guidance becomes a way of giving back when the music business gives you your measure of success, however you define success.)

Reality five

People will disappoint you. Someone you feel you can depend on will disappoint you by not following through with promises or commitments made to you and your artists. Even the smallest oversight by others can have an impact on the things you are trying to achieve for your artist. Anticipate that people will disappoint you, but be pleased when they deliver on their promises. Advice to artist managers: always follow through with your promises because it helps define your character to others.

Reality six

The agendas of many people in the music business determine whether you matter to them. If you are the current manager of a significant artist about to go into the studio to record an album, music publishers will stumble over each other to get the chance for a conversation. If you are the *former* manager of a major artist, you might elicit a faint hello from those same publishers.

This is just the beginning of a larger list, but it covers key points that can easily be taken personally, when in fact, they usually should not be. Recognizing these points as realities of the music business environment can help the manager step away from an issue, realize that it is not personal, and minimize an emotional response to a business situation.

Managers are inconspicuous

Many artist managers are very low key and rarely mentioned or quoted in the press, and so it is important that they actively pursue their own public relations within the music industry. An artist manager's best clients will be the result of the active network of contacts they maintain. This means attending industry parties, events, conventions, and other networking opportunities. Because there is no official Association of Artist Managers in the United States, managers must find events and occasions when they can be seen by being involved socially in the music business on a regular basis. In Britain, however, the Music Manager's Forum (*http://www.themmf.net*) offers regular meetings and seminars locally and around Europe for those employed as artist managers, and it is an excellent way for managers to become part of a network of others in the same profession to help improve their effectiveness on behalf of their clients. Appendix F of this book offers a look at the code of conduct adopted by the Australian Music Managers Forum that is a practical ethics guide for those who manage artists in the music business.

AN UNDERSTANDING OF POWER IN THE MUSIC BUSINESS

In order for the new artist manager to be prepared to navigate the music business, he or she needs to understand where power comes from, who has it, and how it is acquired. Connie Bradley, ASCAP Senior Advisor, says, "Power lies in your friendships. I've never known one person to be powerful without anybody else" (Underwood, 2006). This is a good starting point for a look at where the true power resides in the music business.

Friendships in the music business are much like those in the world of politics: You find out who your true friends are when you no longer hold a position of immediate relevance. True friends quickly step up to offer to help; others who were business "friends" will no longer text, no longer call, no longer email, no longer invite, and no longer return calls. When an artist manager is without major active clients or is downsized from a management company, it can be shocking to learn how weak those relationships with former associates truly are. It can also be reassuring when the manager learns how important the few true friends can be in reigniting a career.

The power of money

As in most industries and in politics, power in the music business comes from having money, the ability to earn money, and money to influence decisions. Major labels, with millions in their annual budgets to promote their projects and those of their partners, are able to claim considerable space on the weekly SoundScan sales charts and on the major airplay charts. It is very difficult for small independent labels to compete at that level.

For example, money shows its power in the music business in the ability of major labels to influence who is nominated, who wins, and who performs on some music awards shows. For some association-sponsored shows, labels actively lobby members of the group and their talent selection committees, and they buy their employees voting memberships in those associations to help influence the outcome of decisions. Money makes it happen. For award shows like the American Music Awards, criteria for an artist's nomination are linked to sales and airplay of recorded music, which virtually assures that only music by the largest labels will be eligible (Serjeant, 2010). Labels whose artists are chosen to perform on awards shows invariably see a short-term boost in sales of their recordings and performance ticket sales, which point to the value of awards shows.

There is an occasional anomaly like the groups OK Go and Straight No Chaser, which can use the Internet with a creative video distributed through YouTube to successfully launch a recorded music project and a tour. However, the power of money at larger labels will continue to help define the major recording artists. To underscore this point, just look at the Top 200 chart in *Billboard* each week to see how very few independent labels can—by themselves—sell at the volume of the majors to get to the top, even in this age of new media.

The power of access

In the music business, the power of access is an important resource for people who have it. Those who can reach key decision makers—or more importantly, influence those individuals—who have power that many do not. This power comes from business relationships and sometimes from associations with others. For example, when Garth Brooks was at the peak of his career, manager Bob Doyle became someone who needed his own personal gatekeepers to filter access to his time. Too many people wanted access to Brooks, and they needed to go through Doyle to reach the artist.

Booking agents have the power of access. Their relationships with prominent concert promoters and other talent buyers can give an artist access to some of the better performance opportunities at the better venues with the more successful promoters.

Radio programmers have the power to give an artist access to their audiences. It is their final decision whether an artist's single is played on their radio station, and that gives them immense control and the accompanying power. With the huge competition among labels for airtime at traditional radio stations, having access to a programmer to promote music on behalf of a record label determines the effectiveness of a label's radio promotion staff. Radio has the power, and the labels invest heavily in the talents of their promotion staffs to access it. Similarly, cable music channel programmers control access to their viewers and have the power to include or reject a music video.

Publicists use their power of access to help their businesses. Those with high-profile clients leverage their association to build on their own client base and to help record labels schedule their artists for appearances on television and cable programs.

The power of your latest success

The old saying that success breeds success also applies to the idea of drawing power from your latest success. Most definitions of success in the music business are based on money. Successful tours, CDs, and songs that find acclaim in their ability to earn give power to those responsible. Producers request the writer's songs, concert promoters seek out the artist's agent, and the artist might be able to renegotiate his or her recording contract for better terms. If your latest success is not followed by another success, there is a strong likelihood that you will become yesterday's news.

The power of your body of work

People with the privilege of an extended career in the music business have built a considerable amount of success. The sheer staying power of individuals who have built long-standing careers within this industry, which disposes of commercially unproductive talent, possess power simply because they continue to succeed. These are people who have the "secrets" to success in the industry and are often sought for their help in opening doors, making introductions, providing advice, being partners, and being mentors. A strong body of work builds reputations, and that translates into power.

Power carries a responsibility to give back

People who have had any measure of success in the music business should use their talents to improve the quality of life of those who need help. The power of music can aid in efforts to raise money for world hunger, for victims of floods or earthquakes, or to pay medical bills for a needy family in a big city or a rural town or village. As artist careers are managed to success, it becomes the responsibility of the manager to seek opportunities for both the artist and the manager to return some of the fruits of their success to the public responsible for helping them achieve it.

THE MANAGER AS AN ENTREPRENEUR

Any person who owns and perhaps also manages a particular business is a *proprietor*. But the individual who has a view for a new business and a new idea of how to make it successful is an *entrepreneur*. The "new idea" element of the definition of an entrepreneur is what sets them apart from other owners or managers of businesses. The artist manager who is an entrepreneur is one who understands the music business, who is willing to make the investment in creating a storefront, and who is willing to take all the risk of possible failure—but also all of the benefits of its success. In a competitive industry that changes as often as it does, artist management in the music business requires success to be built on a continuance of new ideas on behalf of artists' careers.

Earlier we noted that experienced artist managers say that there is a three- to five-year time frame for a new, standalone management company to begin to earn enough income from their artists to become profitable. It takes a lot of confidence, a huge commitment to the vision for the company, and similar confidence in the ability of artists who are signed to be able to develop quickly enough to be able to support the firm. The financial resources needed to develop and sustain a new artist management company are substantial, and so is the courage to accept the risk that goes with starting a new enterprise.

References

Anderson, C. (2006). *The Long Tail*. New York: Hyperion.

Iacocca, L., & Novak, W. (1987). *Iacocca: An Autobiography*. Bantam.

IFPI. (2010, March). *Investing in music*. Retrieved from http://www.ifpi.org.

Kragen, K., & Graham, J. (1994). *Life Is a Contact Sport*. New York: William Morrow and Company.

Leeds, J. (2006). *Personal communication*.

Serjeant, J. R. (2010). Retrieved from http://www.reuters.com/article/idUSTRE69B4TI20101012.

Underwood, R. (2006). Executive Q & A: Connie Bradley. *The Nashville Tennessean* December 10.

The artist: Preparing to be managed

3

Early in the career of an artist, there will be a moment when it feels like it is time to step up the pace of meeting goals and go for the "big time." Some will try to continue to manage themselves, thinking that the 15% commission many managers earn can be saved just by investing a little more time into the business part of being an artist. For artists who seek regional recognition and who record for their own independent label, it is certainly possible to earn a modest income touring and selling their music and merchandise without a manager, although one would be helpful. However, for artists who seek national or international recognition, it is essential to have a manager advocating for them and promoting the growth of their career.

When should an artist seek management? In today's music business world, artists should get a manager when they are prepared to become commercial artists, especially knowing that many deals being made between artists and labels consist of multiple rights or 360 recording contracts. A new artist with most big labels will be offered a contract that requires that the artist must share income streams—ranging between 5% and 30% depending on the contract—so it's important that the artist resist the temptation to sign a recording contract without the guidance of a manager and an attorney. Some artists will feel that they may not get another chance to get a major recording contract and will sign a contract that gives away some of their rights, which might result in them making considerably less money—perhaps nothing—than if they had a manager guiding them in this area. We discuss this more in Chapter 10.

Advocacy by a manager on behalf of an artist is more important today than ever. The International Federation of Phonographic Industry (IFPI) is the counterpart to the Recording Industry Association of America (RIAA). The IFPI reported in 2010 that MySpace featured 2.5 million hip-hop acts, 1.8 million rock acts, 720,000 pop acts, 470,000 punk acts, and 270,000 country acts, as well as millions of acts in other musical categories (IFPI, 2010). To bring these statistics into better focus, in Nashville, Tennessee during a single week in March 2010, the local print media listed 959 different artists appearing at large and small local venues. Each of these artists has a dream, and most are competing for a measure of success in the music business. Without someone to manage their music career professionally, each of the unsigned artists is likely to continue to be a hidden statistic in a data base.

BEING COMMERCIAL IS NOT SELLING OUT

Among the first things that incoming-earning artists must really accept is that they have chosen a career in the music business. It is not a career showcasing the art of music—they already have that part of it figured out. They have elected to become part of the music *business*, and considering themselves to be commercial artists in the music business becomes an acknowledgment that songwriting, performing, and recording are going to be done with the objective of earning money. Many young artists struggle with the idea of becoming commercial until they discover that band members do not want to rehearse without the promise of earnings, and it becomes impossible to keep a performing group together. Likewise, managers are not interested in working with an artist unless there is the likelihood that they can earn money from a music group or individual with a commitment to a career in the music business.

Some young performers feel that becoming commercial is "selling out" their artistry for the sake of money. They resist what they perceive the big label marketing machines will do to their music, because they fear they will be pressured into changing artistically into something that is inconsistent with who they are. However, most labels seek artists who are genuine and unique in their own ways, and who have potential for commercial appeal. Labels sign artists because of who they are artistically and because they think that there is a commercial market for their music, and most want to preserve the uniqueness of their artists that makes them special.

Veteran major label marketing executive Mike Kraski acknowledges, however, that some record labels attempt to "sand off rough edges" from the new artists they sign. In these cases, they may go too far by guiding an artist in a direction, creating "something that is not true to the artist" for the sake of creating something more commercial (Havighurst, 2007). The artist should rely on the strength of his or her manager to insist that the label not homogenize the artist's unique sound.

KNOW WHO YOU ARE ARTISTICALLY

As consumers, we want a commercial product quickly defined in terms that we're familiar with. It is the same in the music business. Artists are asked by key gate-keepers what kind of music they perform. Artists must be prepared to describe what they do musically in very few words. And often an artist's style of music is defined by which radio format would use the artist's music in their programming. When a manager, label, or publisher hears an artist describe his or her music as "contemporary acoustic alternative blues with hints of jam band influences but not as heavy as Hendrix," it suggests that the artist is still trying to find a place in the commercial marketplace and that it may be too early for a management, recording, or publishing contract. On the other hand, when an artist says "my music is alternative rock using catchy lyrics with a killer live show that appeals especially to younger audiences," it shows that the artist has defined his or her music, understands audience types, and knows that appealing to a younger listener is attractive to record labels. In other words, the artist understands that the label is seeking a *business* opportunity through the artist, and the artist is ready to deliver it.

GET EXPERIENCE

Nothing polishes a performer and builds character as much as live performance does. Performing regularly in smaller clubs in front of friends, family, and fans can help artists develop a show in a relatively safe environment. They can take chances by trying new ideas and music to see what is connecting with audiences and what should be cut from the show. Artists can try out new songs they've written or experiment with new material they are considering including in their show. The only warning here is that friends and family can be very supportive and accommodating when they give feedback on performances, and may be inclined to tell the artist what they think the artist *wants* to hear rather than what the artist *should* hear in order to improve their performance.

Live performing coupled with demo recording and songwriting can season an artist and build confidence. Showing an air of confidence without being cocky is a strength managers seek in artists they sign, and that kind of self-assurance can be developed only by getting as much experience as possible on stage, in the studio, and as a writer.

Experience selling tickets and recorded music can be a strong selling point to prospective managers and eventually to record companies. An artist who consistently sells out small venues and sells 6,000 CDs per year at those performances will always get a conversation with a manager seeking talent to manage.

BUILD A NETWORK

A *network* in these terms is a web of supporters with whom artists regularly communicate about their music. An artist's fan base is at the heart of the network. Artists send regular emails to fans to announce new music or new performance dates and locations. They also communicate through their e-teams or street teams promoting themselves and their music, through microblogging services like Twitter, and by keeping their postings to social networking sites current. This continuous communication about the artist—often *by* the artist—keeps fans coming back to online pages and websites, and keeps up the connection with the artist.

Another important part of the network for artists is getting to know those who offer to support their career and to be sponsors when the time comes to push their career to the next level. For example, someone might say something like, "If there is ever anything I can do for you, let me know." That is the cue for the artist to get contact information about the individual, and from that point on that person should be included as a guest at local performances and receive all communications about the activities of the artist. These are the most important fans an artist can develop, because the launch of a career always requires money, and "angels" like these who have offered help can be key to a manager finding funds to develop the artist during those early months of the big commercial push. As a point of reference, other sources of financial support for new artists are savings, family, loans, and sponsorships.

BE PROFESSIONAL

When an artist makes the decision to advance his or her career, it is also a decision to adopt the demeanor of industry professionals from that point forward. That does not mean the artist cannot have fun creating music, but it means that every public performance and industry meeting requires the artist to be completely prepared to make the best possible first impression. Assume that the "lucky break" is in the audience every time there is a performance, and be prepared to deliver the best show possible.

Being professional on the Internet is also important. Amateur sites of artists that are found on the Internet social networking websites are a reflection of how mature the artist has become as a part of the music business, and they also become a statement about the artist's professionalism. Potential career supporters will use an Internet social networking site as a reference point about the artist, and the site should look as professional as the artist can afford. That means artists should seek advice from those who design websites for the music business and then invest a little money to be sure that their very public image on the Internet is one that says they are "somebody." And this is also the time to set up a basic but functional artist website using the artist's name and the domain name as the way to learn about the artist. A *domain name* is merely the location of a website on the Internet, and they can be purchased inexpensively from a number of sources such as Yahoo! and GoDaddy.com.

Using new media also requires the touch of a professional. In an age of sloppy and flippant text messages and careless email drafts, it is important to assume that every message an artist sends will be received by a professional who is expecting the artist to share that same level of professionalism. The use of convenient communication methods does not mean the message writer should not respect the ways these tools are used by business people. Always use good grammar, accurate spelling, and appropriate punctuation when using electronic communication.

BE PREPARED FOR MANAGEMENT

An artist manager will make a number of evaluations about an artist before deciding to offer management, often beginning with a performance at a public venue or a private showcase followed by a somewhat formal meeting to discuss possible management. Certainly the music and preparedness of the artist will be part of those early assessments, but there will be a time when the manager has a conversation with the artist that explores those things that will help make the final determination whether a management deal is practical. Questions that an artist could expect from a potential manager will resemble those someone encounters in any job interview, but an artist needs to keep in mind that a manager is trying to determine the pros and cons in making a decision to manage or not to manage. From the manager's perspective, this decision represents a huge commitment of time to develop the artist, and he or she is trying to find out how many assets—and how much baggage—a potential client brings to the bargaining table. The questions and answers sound formal, because they are. The answers coupled with the SWOT analysis found in Chapter 12 will contribute to the final decision.

1. What do you want to do with your career as an artist, and not just some formal goal—what do you want for yourself that will make you genuinely fulfilled as a creator of music? (The manager is determining what is actually driving the artist's desire for a career and whether his or her expectations of a career in music are realistic.)
2. How much experience have you had writing songs and recording? How often have you performed live over the last couple of years, and did you enjoy it?
3. How are you organized as an artist? Are your band members considered partners or are they hired as needed depending on what the performance requirements are? (If the artist and musicians are part of an organized group, the manager will consider how much income is realistically possible for the group. Income requirements split five ways for a group can be considerably higher than for a single artist who pays band members as an expense.)
4. What do you think is the key to your becoming successful as an artist in the music business? (This is one way to get the artist's opinion on one of their key strengths, but it also lets the manager know whether the artist's view of themselves includes a vision for commercial career in the music business.)

5. Have you ever had an agreement with someone else to manage your career? If so, have you taken steps to formally terminate it? (The manager wants to know if a former manager will show up some day claiming an entitlement to the artist's worth, and to be sure that if one exists, it is taken care of before a management contract is signed.)

6. Do you have your own publishing company and record label?

7. What is it about being an artist that is most satisfying to you? (Over the course of a career in the music business, it will be important to the manager to continue to motivate an artist, and knowing from the beginning where the joy of a career comes from can help to keep the career moving.)

8. If there was one thing about being an artist that you could avoid, what would it be? (The manager wants to know what part of the artist's career development will require work in order to grow.)

9. How much money do you owe? Who do you owe it to? Have you been filing state and federal tax returns regularly? Do you usually pay your bills on time? (Information about finances can show red flags to the manager that will give an idea of what will be necessary to eliminate the distraction of bill collectors. Likewise, it can also reassure the manager that doing business with a financially responsible artist can usually avoid having to deal with finances as an issue.)

10. The last time you had a conflict with someone—a band member, a venue manager, maybe an audience member—how did you resolve it? (The answer to this question will give the manager insight into the emotional maturity of the artist.)

11. If a fan were to describe you as an artist in two sentences, what would that description be? (This tells the manager whether the artist has a realistic view of his or her self-image.)

12. What training or coaching do you think would make you a better performer or songwriter? (This is a back-door approach to asking a question that most people include in an employment interview to get interviewees to list their weaknesses, but it also gives insight into areas that the artist sees as areas of opportunity to advance his or her career.)

13. What is going on in the music business today that you think will be an advantage to you as an artist in the music business? What do you think will be the biggest threat to your career? (The artist's responses to these questions will let the manager know how much the artist has considered the business environment that he or she plans to become part of, and how he or she plans to take advantage of opportunities.)

14. Tell me what you know about our management company. (The manager will use this as a gauge to find out how much—if any—homework artists have done to learn about someone they are considering to direct their career. An artist who knows about the management company is one who probably understands how important the conversation is. If the artist knows little about a management company, he or she is at a serious disadvantage when negotiating a management contract without an entertainment attorney.)

It is clear from these questions and concerns from a prospective artist manager that the artist must be prepared to demonstrate that he or she is ready for the music business. This book is also designed to give artists insight into the artist manager's role in their career and what they should expect from the manager as they prepare to implement a career plan in partnership with the artist. So questions an artist should ask of a prospective manager include:

1. What artists have you managed and what was your success with them?
2. What artists are you currently managing?
3. What do you think of my career possibilities? How do you think I fit into commercial music?
4. How strong is your network—who do you know who you can help you guide my career?
5. What are your expectations on your earnings from my career? What do you expect to be earning commissions from?
6. What expenses will you pay and what expenses will you expect me to pay?
7. Give me an idea what the first six months will be like working with you. How will you use my time to get my career going?
8. Will you be directing my career or do you have others who work for you who will also be involved? What is their background(s) managing an artist?
9. How involved is your management company in client services like publicity and new media?

This is a list that will help artists begin the conversation from their perspective, and it certainly isn't a complete list of discussion points. Some artists will find some of these questions uncomfortable, but the subject of each question deserves a response that the artist should consider. If the management relationship with the artist is to work, it's important that each of these sensitive topics be explored before a formal contract is considered.

PLAN TO BE PATIENT

Most new artists to the commercial side of the music business are part of the echo boomer generation, or people who were born after 1982. These are the children of the baby boomers. Among the attributes of the generation is impatience in achieving success (Leung, 2005). A label executive says that some artists ruin their chances of becoming a commercial success in the music business because they expect too much to happen too quickly. An example he provided was an artist signed to a major label who had five different managers within five years. The relationship with the label becomes strained with so many different people trying to direct the career of an artist. The message here to a new artist is to find a manager who shares your vision for your career, then patiently follow their direction, and become the best business partner you can be.

References

Havighurst, C. (2007, February 7). *All Things Considered: Nashville Band Leaves Label and Thrives*. National Public Radio.

IFPI. (2010, March). *Investing in Music*. Retrieved from http://www.ifpi.org.

Leung, R. (2005, September 4). The Echo Boomers. In *60 Minutes*. CBS Television Network.

Lessons in artist management: From Colonel Parker to Jonnetta Patton

4

Artist management is one of those professions that can be as much art as it is science and business. The artful manipulation of people on behalf of the artist is one of the key functions of the artist manager. Certainly, motivating fans to buy tickets, music, and merchandise are highest on the list of outcomes of good management. Another example is the use of human nature to get past gatekeepers in order to keep an artist's career plan moving. And certainly an understanding of business—specifically the music business—creates an advantage for the successful manager of artists. This chapter draws lessons from the real-world experiences of several veteran artist managers to help build an understanding of this corner of industry.

TOM PARKER

Elvis Presley's lifelong manager acquired the honorary title of "Colonel" as a gesture by a southern governor, but he carried the label with him until his death in 1997. Parker was born in Holland and immigrated to the United States as a young man, working in carnivals and eventually promoting country music shows. It was during his promotion work that he was introduced to Elvis and was asked by his parents to manage the 17-year-old singer.

The first management contract Parker made with Elvis in 1956 awarded him 25% of the artist's overall earnings, but he received half of the earnings from things like recordings and merchandise. Twenty-five percent is a higher commission rate than many management compensation packages provide, but for a new artist without an earnings track record, it is a reasonable rate for the short term. The artist manager will be investing time into the career of a new artist without measurable compensation for a period of time, and receiving a 25% commission when the artist begins generating income helps the manager to recover income from some of that uncompensated time. Several years later Parker renewed his management contract, which awarded him 50% of all of Elvis's earnings and gave him very broad powers of attorney to make contracts on the entertainer's behalf. In his book *Elvis, Inc.*, Sean O'Neal suggests that Elvis did not read contracts of any kind, and he relied on Parker to take care of the details they contained (1996, 75). It was Elvis's inattention to contracts and their provisions that created the opportunity for Parker to, among other things, earn commissions totaling US$1.3 million in 1965, which was actually $300,000 more than Elvis earned that year (11). There are many other stories about Elvis and his relationship with "the Colonel," many detailing similar unconventional business dealings that provided great financial rewards to Parker. Among the best books to recount this history are the O'Neal book and Alanna Nash's book, *The Colonel: The Extraordinary Story of Colonel Tom Parker and Elvis Presley* (2004). With the death of Presley in 1977, Parker went on to help the estate settle its business; however, it was at the continuing rate of 50% commission. Parker died in January 1997 at the age of 87.

Lessons learned

Parker was an adept negotiator. For this era, he was able to get premium rates for Elvis's appearances on television and in movies. He negotiated an adequate royalty rate for his client's first contract with RCA, but later renegotiated contract-tied royalty rates for his recordings to a set amount of money rather than a percentage of the selling price. As the price of recordings increased, Elvis was confined to a set dollar amount for each album sold. This is an example of Parker's lack of understanding of the sometimes-complicated world of the record business, and it ultimately cost him and his client a lot of money. Another example is Parker's failure to register Elvis with a performing rights organization (PRO). Songs that an artist writes or helps to write are entitled to regular payments for the performance of those songs on radio, television, concerts, and other places. Elvis and Parker both were paid nothing because of this omission. To put this into context, in 2003, Clear Channel Vice President Mick Anselmo noted that the five company radio stations in the Minneapolis market paid nearly US$2 million to PROs for the right to broadcast licensed music for the entertainment of their audiences (Anselmo, 2003). With over 10,000 commercial radio stations in the United States, the value of performance payments is considerable for the estate of an icon like Elvis. No one knows how much Elvis missed in performance royalty earnings by failing to fill out a simple registration form with one of the PROs (which are covered later in this book). The inclusion of a skilled entertainment lawyer in all business matters prevents this and other key mistakes on behalf of an artist manager's clients. From the beginning, the artist must have an attorney whose style and manner fits the artist, and the artist and the manager must rely on counsel to help guide the business of the artist's career and to negotiate final contract documents that reflect the best possible financial deal based on where the industry currently is.

RENE ANGELIL: TARGETING

When a family member manages an artist, the opportunities for conflicts of interest are constant. This is especially true when parents manage the careers of their child-artists, as in the cases of Aaron Carter and LeAnn Rimes, who each sued their parents over issues relating to career management.

When that family member is a spouse, however, the results are often positive, as in the case of Narvel Blackstock's management of his wife Reba McEntire's career. Another example is Rene Angelil, husband and manager for Celine Dion and who is 26 years her senior. Angelil began his career as an artist manager, following his work as a member of a Canadian group called the Baronets. His group built a reputation performing in Quebec in the 1960s (Charlebois, 2004). His career transition to artist management ultimately linked him in the early 1980s with 12-year-old Celine Dion. She had sent an unsolicited recording to Angelil with a request that he consider managing her career. He was slow to respond to her request, but he finally met with her and signed her to a management deal (Proefrock, 2005). Angelil believed in his new artist enough to mortgage his own personal assets to finance two albums for her, both of which generated considerable attention for the young performer. Angelil signed a new contract with Dion when she became 18, and both agreed that his commission would be an unusually high 50% (Baunoyer, Beaulne, and Wilson, 2004). Commission is a percentage of an artist's income paid to a manager, which typically ranges from 10–25%, usually in the lower end of the range.

As Canadian-born Dion pressed Angelil to make her an international star, he knew that an image makeover would be necessary for the French-speaking singer. Among the changes he made: sending her to school to become more fluent in the English language, ordering a general makeover that included cosmetic dental work, and having her hair restyled. The result was a launch into the lucrative American music market that garnered her Grammy awards, helped her sell over 50 million of albums in the United States, and made her one of the biggest acts to ever perform in Las Vegas.

Now in his late sixties, Angelil continues to manage his wife's career despite his battles with heart attacks, cancer, and lawsuits from those who know about his deep pockets.

Lesson learned

An artist manager must have a keen sense of the target market for a recording artist. In its strictest sense, a *target market* is defined as people who are able to and willing to buy concert tickets, recorded music, and merchandise. Dion's hopes were to enter the American marketplace and become an international star. Angelil knew that her success in the United States would require her to have a better command of the English language so that she could effectively communicate her art through the American media. Her image makeover and new language skills were among the keys to her commercial success in the United States.

MICHAEL JEFFREYS: CONFLICTS OF INTEREST

Conflicts of interest are business relationships people have that may cause their decisions to be made on their own behalf rather than for the benefit of the person or company they represent. For example, an artist manager is in a position of trust with the artist, and would have a conflict of interest if he or she owns a recording studio to which the manager might direct the client. He or she has a personal financial interest in a business that may compete with the professional interests of his or her client. To effectively manage an artist, there is a requirement that the business side of the relationship should be unencumbered by agreements that cause outright conflicts of interest. The managers of rock icon Jimi Hendrix had some conflicts of interest, and the result shows the negative impact on the artist's career.

Michael Jeffreys and Chas Chandler signed a co-management contract with Hendrix in 1966. They agreed to manage the career of the guitarist for 30% of his earnings, which was considerably higher than the standard 15% charged by many artist managers. Added to that was a 3% product royalty that the managers received on all of the Hendrix recordings, and they earned half of the money generated by a co-owned music publishing company. In all, the Hendrix co-managers' earnings from the artist were considerably higher than typical management agreements (Hopkins, 1984).

Two years after Jeffreys and Chandler agreed to manage Hendrix, Chandler wanted out of the arrangement. Jeffreys bought Chandler's portion of the contract, with the result that he began earning as much as 40% of the artist's income. With Hendrix as Jeffreys' primary client and only source of income, Jeffreys began to book shows for the artist on an intense schedule, often using his personal requirements for income rather than making sensible decisions based on Hendrix's career and the artist's capacity to deliver top performances at each concert stop (McDermott, 1992).

In September 1970 Hendrix died, ending a remarkable and short career, and setting to rest an artist-manager relationship that had its difficult times. In 1973, Michael Jeffreys died in a plane crash in France, with his artist management assets passing to his father. Michael Jeffrey's father Frank settled the estate of his son, and it was then that the extent of the conflicts of interest became very public. The publishing and recording contract Jeffreys and Chandler had with Hendrix said, in part, "All recordings made hereunder and all records and reproductions made therefrom, together with the performances embodied therein, shall be entirely our property, free of any claims whatsoever by you or any person deriving any rights or interested from you." It remains an unusual feature of an artist-manager contract that the manager becomes owner of the artistic creations of those they manage (Goodman, 2004).

Lesson learned

The artist–manager relationship is built in part on trust, and the expectation is that the manager will make decisions that are best for the artist's career. In the end, good decisions on behalf of the artist will result in continued long-term financial gain for both the artist and the manager. To put it another way, the agenda of the artist must come first. In the case of the Jeffries–Hendrix relationship, from the very beginning the manager was drawing more from the income stream of the artist than is customary. An argument could be made that earning a high percentage of Hendrix's income was Jeffries' incentive to focus on making his artist a megastar. However, the amount of money generated by the artist's career perhaps made Jeffries unwilling to develop other acts, and Hendrix was the sole support of his manager's lifestyle. Decisions by Jeffries became driven by how much money he could make for himself by pushing Hendrix's career at a pace that became physically and emotionally taxing for the artist. Jeffries owned song publishing, recordings, royalties earned by the recordings, and a recording studio that would be used exclusively by Hendrix. All of those conflicts of interest have the potential to cloud the objectivity of managers to impartially guide the career of an artist in the music business.

PETER GRANT: A SHARED BELIEF BETWEEN THE ARTIST AND THE MANAGER

In an industry that is filled with hyped images and exaggerated perceptions spawned by the nature of the entertainment industry, a manager of a performing recording artist must genuinely believe in the artist for who they are as an artist. Probably there has been no deeper belief and commitment to an artist than Peter Grant had for Led Zeppelin.

Grant was born in England and became a factory worker, a photographer, a waiter, a professional wrestler, and a stagehand. By his mid-twenties, he was driving American bands to London-area performances, where he became somewhat familiar with the general workings of performing acts. Promoter Don Arden, Sharon Osbourne's father, hired Grant to become the tour manager for American artists like Little Richard and the Animals. He became adept at tending to the affairs of artists performing on the road because of his experience, and in part because of his large presence. He stood 6'6" tall and weighed well over 250 pounds. His imposing presence and the knowledge that he occasionally carried a gun made him a natural to create order out of the chaos that sometimes accompanies touring (Davis, 2004).

Grant started a management company with friend Mickie Most and acquired the Yardbirds as one of their acts. Eventually Grant bought out Most's portion of the Yardbirds management agreement and became their sole manager. The Yardbirds' band members varied over time; at times, Eric Clapton, Jeff Beck, and Jimmy Page were each members. As their manager, Grant's experience in road management proved to be the key to making the Yardbirds' concert tours profitable after months of losing money (Welch, 2002).

When the Yardbirds broke up, Grant formed a new group with Page and some new band members, calling them the New Yardbirds. This group morphed into the legendary Led Zeppelin. Under Grant's guidance, the group signed a contract with Atlantic Records that featured a $200,000 advance (one million U.S. dollars in today's value) and full control over writing, publishing, and recording. In an era when concert promoters were receiving 40–50% of the earnings from a ticketed performance, Grant negotiated a 90/10 split with promoters, with Led Zeppelin receiving a record-level share of gates from performing. With the huge success of the group as a touring band, he convinced promoters that a 10% cut of the concert was more than adequate given that the group was making (in current dollars) a million dollars for every show date in 1973 (York, 1993, 299–300). Grant's decision to limit Led Zeppelin's exposure to the media and to seek album sales rather than the sale of singles was among his nontraditional approaches to artist management in the 1970s. In 1980, the death of a member of the group (John Bonham) led to the end of Led Zeppelin. In 1995, Grant died at the age of 60 from a heart attack (Clark-Meads, 1995).

Lesson learned

The most effective manager is one whose belief in the artist is deep enough to be the basis for every decision made on his or her behalf, whether it is believing in the artist's potential or believing in who the artist is. Peter Grant was constantly on tour with Led Zeppelin, handling most of the tasks associated with tour management and artist management. But when the time came to create the words and music, and to assemble the performance, he left these creative responsibilities in the hands of the group. Likewise, the band left the management decisions—including some very unconventional ones—up to Grant. This shared and deep belief in each other became what many acknowledge as one of the strongest bonds between artists and a manager in the music business.

HERBERT BRESLIN: PROMOTING YOUR ARTIST

Among the critical skills for the manager of an artist in the music business is the ability to find all of an artist's artistic appeal and to then know how to promote and sell it to buyers of tickets and music. The most effective managers have the knack of knowing which opportunities are best for their clients and how to turn them into promotional events to build interest in the artist and grow a developing fan base.

Long before he met the late Luciano Pavarotti, New York–born Herbert Breslin was introduced to opera as a young boy by his father. He learned to save his money in order to buy tickets to see events at the opera house. Later he joined the U.S. Army, and after leaving the military, he earned degrees from City University in New York and Columbia University. Among his continuing interests was the opera, and he accepted a nonpaying job handling press and publicity for the new Santa Fe Opera in New Mexico. He developed his profession with the company and went on to establish his own career as a public relations manager and publicist for classical music.

In 1967 he met Pavarotti, which began a 36-year professional relationship between the artist and his manager. Initially, Breslin was employed by the opera singer to handle publicity and public relations for him, but later began to handle most of the traditional management duties of a performing artist. For his work, Breslin was paid 10% of Pavarotti's earnings, and according to the manager, he worked for the singer without a contract (Breslin, 2005).

Breslin's ability to connect his opera star with the media is fabled. He worked well with the media and would create six or seven events within a short period of time to highlight a tour or to promote a new album. Although Breslin has his critics, few deny his ability to understand his clients and to know what it would take to promote and manage their careers. He is now retired from management.

Lesson learned

An artist manager should draw from the strengths of artist clients and use them to add as many dimensions as possible to their persona and their commercial appeal. The phenomenal caliber of the late Pavarotti's voice is undeniable; it was Breslin's work using his singer's voice and warm personality that broadened his commercial appeal beyond the opera stage. An important talent of a manager is understanding career promotion and having access to the tools to make it happen. In the case of Breslin and Pavarotti, the manager says of his former client that he has "a great face, a wonderful smile, and a wonderful sense of humor. He charmed everybody"

(Breslin, 2005, 65). And it was Breslin's exploitation of these assets that helped greatly expand the career appeal of his most famous client.

JOE SIMPSON: MANAGE BY THE BOY SCOUT MOTTO

Joe Simpson has been a psychologist, a youth minister, a record producer, a television producer, and is the father of Ashlee and Jessica Simpson (MTV, 2006). It is perhaps his experience as a psychologist that most prepared him to manage and promote these two young women in the music business.

Simpson tried to help Jessica launch a career in Christian music in the mid-1990s, but her music never connected with a public that buys recorded music. In 1997, Sony received a demo of her music and signed her to a recording contract that resulted in a very successful first album—her career was then well underway under the guidance of her father. Her music continued for two more albums, but her popularity as a recording artist was beginning to fade. Joe Simpson decided to try to sell the idea of a reality show featuring newlyweds Jessica Simpson and Nick Lachey. The show reignited interest in Jessica, and he was able in 2005 to leverage this interest to get his daughter a major role in the movie version of *The Dukes of Hazard*, but her recording career withered in the following years, most recently with a failed attempt at country music in 2008 and 2009.

Joe Simpson also planned and eventually launched the music career of his younger daughter, Ashlee, in 2004. She too had two early and successful albums and a reality television show, but her recording career also went into decline with her most recent album in 2008 barely selling 100,000 copies.

With music performance as the foundation for the careers of Simpson's two daughters and with him as advisor for opportunities to advance their careers, both women fell into circumstances that drew criticism. In 2004, daughter Ashlee was caught on live television—*Saturday Night Live*—using a prerecorded track of her voice (rather than her live voice) that didn't sync with her live band. She walked off the stage blaming her band, while her record label blamed a computer problem. In late 2006, daughter Jessica performed a song at a taped television salute to the career of Dolly Parton, but her performance was so bad that it had to be cut from the version that was broadcast (Moraes, 2006). These two very public career gaffes made them the subject of jokes on television and headlines in the tabloids.

Lesson learned

A manager cannot overprepare an artist for a public appearance. Mistakes like those made by the Simpson sisters can erode their images as professionals and make their fans uneasy about telling others that they are admirers, which can be fatal for a career. Be sure that the artist is comfortable with performance opportunities, and then be sure that they are prepared.

JON LANDAU: KEEPING A BUSINESS FOCUS

Jon Landau's relationship with Bruce Springsteen spans more than 30 years. His background in the music business included being an editor for a major music industry publication, a film critic, a music producer, and—most famously—an artist manager.

His connection with Springsteen began when he attended one of Springsteen's New York City shows in 1974. The two became friends and Landau became a valued influence on Springsteen's work in the recording studio. The closeness of the relationship created conflict with Springsteen's manager at the time; Landau became his manager and has guided his career since then.

Springsteen always considered himself an artist and shied away from the commercial aspects of performing music. Part of this was attributed to the mistakes made selling merchandise at performance venues early in his career. However, Landau was able to show his client over the years that a career is based on the continued success of the business it generates. Springsteen began to accept that. Landau has a style of communication that made his client comfortable with the idea that his music can speak for working people, yet still have a complementary commercial aspect. Springsteen continues to sell platinum albums, and his 2009 tour grossed over US$95 million (*Pollstar*, 2010).

Lessons learned

Many artists launch their careers with a vow not to "sell out," meaning that they will shun any attempt by those around them to convince them to become commercial. They say that they want to be true to their art and true to their fans. Commercial music, they often say, is cookie cutter music, and they are unwilling to become something they are not. An effective artist manager can guide the artists in the business side of their craft and still show them the way to reach a larger fan base. When artists say it is time to have a manager, they are also saying that they are at a point in their career where they want their artistry to begin building a financial future for them. It's not necessary to compromise values to reach commercial markets that buy music and performance tickets. Rather, it takes the guidance of a manager who knows how to expose those creative features of an artist that appeal to a larger audience. Jon Landau had the industry background to understand the business and the human relations savvy to work with his artist, showing him that you don't need to sell out your artistry in order to have financial success.

BOB DOYLE: USING YOUR NETWORK

Bob Doyle is a product of Nashville's part of the music business. He worked for Warner Bros. Records in the A&R department, at ASCAP in Nashville overseeing member relations. In 1988, he left the performance rights organization to open the doors of his new music publishing business, Major Bob Music. One of his first clients was a talented singer–songwriter named Garth Brooks.

He used his connections to try to find Brooks a manager, but no one was willing to take on a new client. So Doyle decided to do it himself. He recruited a new partner, publicist Pam Lewis, to help him manage Brooks. The backgrounds of Doyle and Lewis meant that they had a number of connections necessary to get Brooks' career moving. Doyle put Brooks together with producer Jerry Kennedy to create a music demo. Kennedy was then able to connect the artist with booking agent Joe Harris of Buddy Lee Attractions. Doyle used his contacts to pitch Brooks to every label in Nashville but failed to get any interest in his client (Morris, 1992). In May 1988, Brooks was scheduled to appear at the legendary Blue Bird Café in Nashville to showcase some of his songs. The lineup and performance times for the show were changed at the last minute, and he found himself performing before several record label chiefs, with the result that he was signed by Capitol Records (Mitchell, 1993).

The career of Garth Brooks continued to be managed by Doyle through the best of times an artist can experience. Along the way, Doyle and Lewis parted, but Doyle continues to manage the career of one of the most commercially successful artists in history. His company is now known as Bob Doyle & Associates and his firm manages other artists and he continues to manage Garth Brooks.

Lessons learned

Bob Doyle was able to move Garth Brooks through the maze of the music business in part because he knew the key gatekeepers—or he knew those who knew them. Inexperienced artist managers find it difficult to make contact with key people in the music business because they don't have a network of relationships.

The industry has hundreds of thousands of musically inclined individuals with lofty dreams of stardom, and all are trying to connect with those who can make a career happen. The result is that offices of management companies from Los Angeles to Nashville and from New York to London are bombarded daily by those who need help realizing their dream in the music business. Artist managers new to the industry usually don't have the connections—or network—to get telephone calls returned on behalf of the artists they want to manage. Bob Doyle and Pam Lewis already had access to key people in the music business, which added to their effectiveness working on behalf of Garth Brooks' career.

ANDREW LOOG OLDHAM: EXPLOIT YOUR ARTIST'S TALENTS

Nineteen-year-old Andrew Loog Oldham recruited and signed the Rolling Stones as his first act to manage. His background was limited but impressive. He had been an assistant to Brian Epstein, the manager of the Beatles.

His business savvy was quick to show. He signed his management deal with the Stones on May 1, 1963. Nine days later, the group was recording its first session for Decca Records (Cannon, 1992). Oldham tells journalist Sean Egan that he approached the head of Decca A&R, Dick Rowe, and pitched the Rolling Stones, knowing that Rowe had been offered the Beatles but had declined to sign them. Oldham knew that Rowe didn't want to be known as the guy who passed on the Rolling Stones, too, so he got his deal (Egan, 2001).

Oldham knew that singing songs written by others would not set the group apart as much as writing their own original material for their album projects. He pressed Keith Richards and Mick Jagger to begin writing songs, although Jagger wasn't a musician and Richards had done little songwriting. The result was that some of the most remembered music of the 1960s was penned by these two Rolling Stones, including "(I Can't Get No) Satisfaction."

Lessons learned

One of the roles of an artist manager is to exploit all of the talent that artists have. If an artist hasn't written songs, the manager should begin developing that side of the artist. Although it is possible that the artist will never become a prolific writer, it will give him or her some insight into the creative processes of others who bring their songs to the studio. If artists have a creative side that permits them to write commercially viable songs, the manager has done his or her job by stimulating a creative revenue source for the artist.

JOHNNY WRIGHT: A MATTER OF TIMING

Johnny Wright is an artist manager who has also been part of the management teams of some of the most commercially successful pop music acts in history. His resume includes work with Britney Spears, the Backstreet Boys, 'N Sync, the New Kids On the Block, Boyz II Men, Justin Timberlake, the Jonas Brothers, and Sean "P Diddy" Combs. He began his association with the music industry as a radio DJ, and was recruited to travel with the New Kids for over four years. During those years, Wright learned a considerable amount about artist management and the challenges that go with managing artists.

Being on the road with a performing act exposes tour managers and aspiring artist managers to most of the issues one will face in the profession. Any tour, whether it is one that takes the artist and crew out for six weeks or one that hubs (returning home after each performance date), includes continuous contact with tour promoters, radio stations, venues and venue management, talent bookers, lighting companies, video companies, sound companies, production coordinators, the entertainment press, musician and stagehand unions, sponsors, fans, security companies, local/city governments, hotels, caterers, airlines and other transportation companies, and many others we discuss later in this book. All of these entities can create course of action that needs correcting, and being on the road and participating in solutions is among the best ways to learn how to manage issues.

Lessons Learned

Johnny Wright was able to develop his expertise in artist management for the music business early in his career, in part because he was able to place himself into the role of problem solver when many of the problems occurred. Because most issues are related to people and how they handle their responsibilities, they all have a human element, meaning that each situation is as different as the people who are part of the circumstance. Learning how to motivate people to assure the success of a performing artist is a critical component of the profession of artist management.

LOU PEARLMAN: A MATTER OF TRUST

Lou Pearlman is another artist manager who began in a different career but ultimately became a successful career manager: in this case, for the Backstreet Boys and 'N Sync. Pearlman was raised in Flushing, New York. Among the things he did in his youth was becoming a musician in a local band. He eventually found his way into the aviation industry, where he was a very successful entrepreneur owner of a helicopter and aircraft charter service. His charters provided services to several key music industry people and this—coupled with the fact that his cousin is Art Garfunkel—opened a network of contacts that helped him become a major player in the music business. He put together the elements that resulted in the creation of the Backstreet Boys, and he later signed 'N Sync to a management contract.

One of the continuing truths of nearly any industry is that the wealthiest are always prone to being sued. In some cases, it is because the wealthy have "deep pockets" and people sue for some of what they have. In other cases, the wealthy leave themselves exposed to being sued. This was the case between Pearlman and the Backstreet Boys. The group had signed a management contract with Pearlman when some of the group members were barely teenagers, and court documents listed among the things Pearlman was accused of by the five Backstreet Boys:

- Pearlman made himself the sixth member of the group, which entitled him to a performer's percentage of the earnings from touring and merchandise sales.
- Pearlman paid his own company, Trans Continental, 43% of the earnings of the group as consulting fees.
- Pearlman booked all of the group's flights on his own airline charter service.
- Pearlman put the savings of the group members into investment savings accounts owned by his own company (Schneider, 2000).

These issues were settled out of court.

Lessons learned

Among the key elements of the artist manager relationship is trust; this relationship is often compared to marriage. Some marriages begin with a prenuptial agreement; likewise, an artist–manager agreement begins with a contract. Marriages include the earning of income shared by the partners, and an artist shares income with a manager. Managers are often given power of attorney, which lets them act on behalf of the artist under certain circumstances; likewise, marriage partners frequently and unilaterally obligate the partnership. By contemporary standards, it could appear that Pearlman took advantage of the Backstreet Boys. However, it is unclear how much of his personal resources was invested into the group and it is equally unclear what the group understood would be the basis for repayment of Pearlman's investment and his continued earnings. When the Backstreet Boys—now young men—realized what the costs were for Pearlman's management, they felt they had been taken advantage of and sued him. When such a relationship loses its foundation of trust, it is all but impossible to recover from it.[1]

SHARON OSBOURNE: A FAMILY EXPERIENCE

Sharon Rachel Arden is the daughter of the original manager of the rock group Black Sabbath. Ozzy Osbourne began the major part of his career as part of the group until he was fired and began to work solo. During these early years, Sharon worked for her father in his management company, gaining the experience necessary to later launch her own career as Ozzy's manager. Sharon sought to be Ozzy's manager as part of the Arden management company, but her father refused. So she bought the management contract from him, began to manage Ozzy, and along the way became Mrs. Osbourne. Sharon negotiated Ozzy's solo recording contract with Epic in 1981; his first solo album sold over 4 million albums (Rosen, 2003). She continues to manage all aspects of his career, which included a top ten album in 2010 and career music sales of over 20 million albums.

Lessons learned

Among the best lessons that artist managers can take from Sharon Osbourne are those she shared about her experience in *Fortune* magazine. In the May 13, 2002, edition of *Fortune*, Alynda Wheat wrote about Sharon Osbourne's philosophy on artist management based on her experiences with Ozzy Osbourne. Her major points are:

- Manage Your PR
- Marry Your Job
- Customer Satisfaction Sells
- Follow Your Gut
- Play to Win
- Learn to Cope
- Know When to Move On

---

[1]Pearlman was sentenced in 2008 to 25 years in prison for defrauding investors of hundreds of millions of dollars in a Ponzi scheme.

She uses these points as a basis to briefly discuss, in classic Sharon Osbourne style, her career experiences managing Ozzy Osbourne. I urge the interested reader to find a copy of the issue for the complete article.

LUKASZ GOTTWALD (DR. LUKE): HAVING IT IN WRITING

Dr. Luke was producing Kelly Clarkson songs in 2005, when he received a song demo from a friend at BMI and became interested in the singer on the tracks, Kesha Rose Sebert, later to become Ke$ha. The demos by the 18-year-old impressed Luke, and he signed her to his own label, Kemosabe Records, with the intention of producing and releasing her music when he was able to find time in his schedule. For management, Ke$ha signed a contract on January 27, 2006, with DAS management, which is reproduced in Appendix E of this book. The agreement between DAS's David Sonenberg and Ke$ha contains most of the standard provisions of a management contract, but it also includes a provision that she may end the contract if DAS doesn't have a distribution deal with a major label within a year of her signing the management agreement with DAS. By 2008, Ke$ha didn't have the promised major label recording contract, although Warner Bros had expressed interest in signing the artist; however, the label ultimately backed away from the deal because of the existing agreement with Luke's record label. DAS and Ke$ha parted ways.

Luke began managing her career and guided her to RCA, where she received a recording contract in 2009 and began working on her first album, "Animal." The album was released in early 2010, and the music from it set chart and sales records. Her first single, "TiK ToK," became the longest-running single on the *Billboard* Hot 100 chart by a female artist in 33 years. SoundScan reported sales of the song at 610,000, making it the best weekly sales on the digital song chart ever by a female artist. Her album went well past platinum status, and she has sold over 11 million digital singles since the release of the album (Caulfield, 2010).

With Ke$ha's enormous commercial success, it came as no surprise that Sonenberg sued Luke, seeking millions in damages from him and claiming that he had convinced Ke$ha to terminate the DAS contract and to disavow the Warner Bros. contract that Sonenberg's company had negotiated on her behalf (see Appendix E).

Lessons learned

As a manager, it is important to protect the time that you invest in the development of an artist with a reasonable contract. The contractural promise of a major career step within only 12 months of signing an artist management agreement is risky, even with the most talented new artist a company acquires. As you will see later in this book, developing the career of a new artist can pay dividends for managers who successfully direct a commercial music career, but those dividends typically require years of work without significant repayment for the time invested until the artist connects with their market.

JONNETTA PATTON: MANAGING AN ARTIST'S IMAGE

Usher Raymond IV grew up in Chattanooga, Tennessee, singing in the choir of a Baptist church that he attended with his mother, Jonnetta Patton. His mother recognized his gifts of musical talent and a charming personality, and took on the responsibility of managing a potential music career for him before he was ten years old. She placed him in a boy band called New Beginnings when he was 12, but later entered him solo in a talent show in Atlanta, which resulted in a record deal with La Face records. The label released a single by Usher on the *Poetic Justice* sound track. His voice began to change when he was 15 and he nearly lost his recording contract. To prove that Usher was ready for the next stage of his career, in 1995 his mother/manager put him on the road playing increasingly larger clubs and building a large fan base. His music and his "ladies' man" image proved enough for an album, *My Way*, which sold over 4 million copies.

On the strength of music video images of Usher as a single, partying, sex symbol, his next album, *Confessions*, sold over 14 million copies, and he became one of the biggest music stars of 2005. During the next four years, Usher was married, had two sons, got divorced, and in 2008 released an album that had—by Usher standards—a lukewarm reception. Many pointed to their difficulty accepting the new image of Usher in a domestic role—a genuine collision of perception with the reality of Usher the person. His public image had gotten away from him. Several accounts acknowledge that his mother disapproved of the relationship that became a marriage, and it caused a strain on her personal and business relationship with her son. She was out of her son's career in 2007 and then back into it again in 2008, as he tried to recapture his career status and the old image again. Two single platinum albums followed, but none as strong as the ladies' man success story in 2004.

Lessons learned

The easy lesson to draw from this could be that a manager must do what's necessary to preserve an artist's image. But that would be the easy way out. Rather, it is important to manage an artist's image as a career and the years mature him or her as a personality and a performer. The tastes and musical preferences found in pop culture move too quickly, and a manager must work to be sure the artist continues to adapt and meet expectations of fans.

An example of someone who has worked well with his artist's maturing image is Paul Rosenberg, Eminem's long-time manager. In 1999 Slim Shady was on his way to becoming a rap icon, dressed in the obligatory ball cap and baggy shirt and pants and sporting blond close-cropped hair. Through the years, he added quite a few tattoos and lost the blond hair. But by 2011, the tattoos were covered, the bling was minimal, and the 37-year-old's stage image was mature, but his music and performances were still all message and attitude. By this time, many from his 1999 fan base were approaching 30 years old and Rosenberg and Eminem adapted the presentation of the music to meet the expectations of his newer and his older fans, resulting in his album *Recovery* topping sales charts for eight weeks in the summer of 2010, which was on top of the more than 40 million albums he sold in the United States over the course of his career.

References

ABC News. (2004, October 29). *Pop Stars Jessica and Ashlee Are Pop's Stars.* ABC News Internet Ventures. Retrieved from http://www.abcnews.go.com/2020.

Anselmo, M. (2003, November 4). *CRB Fall Forum.* Nashville.

Baunoyer, J., Beaulne, J., & Wilson, D. (2004). *Rene Angelil: The Making of Celine Dion.* Dundurn Group, Toronto.

Breslin, H. (2005). *The King and I,* Mainstream Publishing.

Caulfield, K. (2010, February 27). *Animal Planet* (p. 23). Billboard/BPI Communications.

Cannon, B. (1992, March 6). *Casting the First Stones.* Entertainment Weekly.

Charlebois, G. (2004). Would that Be Angelil at the Top of the Food Chain? *Montreal Quebec Gazette.*

Clark-Meads, J. (1995). *Groundbreaking Manager Peter Grant Dies at 60* (p. 12). Billboard/BPI Communications.

Cohen, J. (2006). *Springsteen Does Seeger on First Covers Album.* Retrieved from http://Billboard.biz.

Davis, S. (2004). *Hammer of the Gods.* Berkley Boulevard, New York.

Egan, S. (2001, December 26). Oldham Still 'Rolling' with the Stones. *Billboard.*

Goodman, C. (2004). UK Charities in Battle for Jimi's Millions. *Sunday Express.*

Hopkins, J. (1984). *The Jimi Hendrix Story.* Sphere Books, UK.

Jimihendrix.com. (2006).

Lebrecht, N. (2004). The Man Who Made Pavarotti Very, Very Rich. *The Evening Standard,* story available at http://gulfnews.com/life-style/people/the-man-who-made-pavarotti-very-very-rich-1.337829.

McDermott, J., & Kramer, E. (1992). *Hendrix: Setting the Story Straight.* Warner Books, Inc, NY.

Mitchell, R. (1993). *Garth Brooks: One of a Kind Working on a Full House.* Fireside.

de Moraes, L. (2006, December 22). Star-Stricken Jessica Simpson Undoes the Honors for Dolly Parton. *Washington Post,* section C, 1.

Morris, E. (1992). *Garth Brooks: Platinum Cowboy.* St. Martin's Griffin, NY.

MTV.com. (2006). *On Air.* MTV Networks.

Nash, A. (2004). *The Colonel: The Extraordinary Story of Colonel Tom Parker and Elvis Presley.* Chicago Review Press.

O'Neal, S. (1996). *Prima Lifestyles.* Elvis, Inc.

Pollstar, (2010, January). *Chart of top tours of 2009.*

Pearlman, L. (2003). *Bands, Brands, & Billions.* McGraw-Hill.

Proefrock, S. (2005). *All Music Guide.* Retrieved from http://vh1.com.

Rosen, C. (2003). Spotlight: The Osbournes—Sharon Osbourne's Tragedy and Triumph. *Billboard.*

SoundScan. (2010, Spring and Fall).

VH1. (2010, July 20). *Behind the Music: Usher.* Retrieved from http://www.vh1.com/video/shows/full-episodes/usher/1643930/playlist.jhtml.

Welch, C. (2002). *Peter Grant: The Man Who Led Zeppelin.* Omnibus Press.

Werde, B. (2010, February). *Golden Girl* (p. 21). Billboard/BPI Communications.

Wheat, A. (2002, May 13). How to &#$@ Manage Like Sharon Osbourne. *Fortune.*

York, R. (1993). *Led Zeppelin: The Definitive Biography.* Underwood Miller.

The artist management contract

5

When the artist and the manager are ready to shake hands and begin their business relationship, it means they have agreed in principle to a contract and are prepared to exchange promises. Artists agree to have their careers managed and by no one else, and the manager agrees to manage artists' careers and exploit their talents for commercial gain. That is the short version. As in all business matters, agreements are formalized within a contract in order to have a record of what responsibilities each has pledged to the relationship. This chapter is not so much about law, but rather about reviewing key provisions of a contract between the artist and their manager and understanding the importance of those provisions in defining the relationship between the two. Although there are some artist–manager relationships that are not memorialized with a written contract, something as brief as a written management summary can provide at least a measure of understanding of the promises the artist and the manager are exchanging.

NEGOTIATING THE CONTRACT

Before we consider contract elements, it is important to understand that in this setting the artist and the manager employ separate attorneys. There isn't a case in which a single attorney can represent both parties to a contract without creating a conflict of interest. This is a business transaction involving negotiated provisions for two separate parties. By the nature of the transaction, it is required that each party has an attorney to represent his or her interests regarding the terms of the agreement. Each must have separate legal representation.

THE LENGTH OF THE CONTRACT

Artist management contracts approach the term, or length, of a contract in two common ways. The artist will often agree to a three-year management term and offer the manager the option to extend the contract an additional two to four years, provided that both agree to the extension. The manager who chooses not to extend the contract should give the artist adequate notice that they don't wish to continue to manage the artist beyond the expiration of the current contract. In traditional business settings, a two-week notice is all that is expected when an employee decides to end employment; however, the complexities of managing the many facets of an artist's career will require a considerably longer notice of intent not to extend the relationship.

An alternative to a management contract defined by time is one that is linked to a certain number of albums. For example, the artist might seek someone who will agree to manage their career for two album cycles with the option to extend for one or two more cycles. The top 200 albums listed from the Billboard 200 chart in September 2010 shows the average weeks on the chart to be nearly 24 (SoundScan, 2010). Considering the amount of time necessary to launch an album and ride the wave into a second album, it is reasonable to predict that a management agreement predicated on two album cycles could easily be for a time period of two years or longer if the artist is active on the sales chart. Album cycles differ among genres. For example, successful country albums tend to have a much longer cycle than successful hip-hop albums. Keep in mind that many commercial albums release a single from a new album six to twelve weeks before the full album is released, and an artist will require considerable assistance from a manager to plan the activities necessary to support the album through public appearances, performances, and other promotional activities.

The method that is used to determine the length of the contract will in part depend on the attorney giving the advice. Some attorneys prefer their contracts to include finite terms that indicate specific amounts of time like months or years; others are comfortable recommending the less specific time restrictions of album cycles to determine when contracts end. Which method is chosen has a lot to do with the experience of the manager and the maturity of the artist's career.

From artists' perspective, having a shorter term with an option to extend an agreement will provide a safety net allowing them to end a relationship with a manager that ultimately may not work out. On the other hand, the manager who takes on a new client with an undeveloped career may wish a longer contract in order to recoup anticipated losses and foregone income that inevitably comes with the decision to manage a budding artist.

THE MANAGER'S SERVICES TO THE ARTIST

This section of the artist manager's contract is very much like any job description you will find. It mentions the specific things the manager is to do on behalf of the artist; these contracts often include a catch-all phrase similar to what everyone finds in a job description: "The manager will do whatever else is necessary to effectively advance the career of the artist."

Specifically, the management contract lists those advisory duties of the manager, which include providing advice on:

- All phases of their career in the entertainment industry
- The appropriate music and show design for live performances
- Publicity, public relations, employment, and advertising
- Image and related matters
- Booking and talent agencies that work on behalf of the artist
- The selection of other key team members, such as attorneys, business managers, accountants, publicists, and webmasters

This section of the contract also requires that the manager do anything else that could reasonably be expected of an artist manager. Keeping this requirement relatively open-ended takes into account the rapid changes made by technology and evolving business model of the music business.

Another key component of this section of the artist's contract is one that specifically says that the manager isn't obligated "to seek, obtain, or procure any employment or engagements for the artist." The reasoning behind the need for this section requires explanation. The artist's booking agent has the responsibility to secure performance opportunities and contracts, although the manager often assists in developing the contacts necessary to get those deals done and provides final approval for performance agreements. California law is very specific about what an agent does. In order to be a booking agent who negotiates employment contracts involving California businesses (clubs, venues, etc.), an individual is required by law to be licensed according to the California Talent Agency Act, Labor Code section 1700.4. The state labor commissioner issues a license to those who book shows, and failure to be licensed is a crime. A similar law applies in the state of New York.

These and similar laws are designed to prevent managers of artists from being paid a double commission—one from the artist's general earnings and an additional commission as the artist's booking agent—which effectively creates a conflict of interest. For example, a manager who books a $10,000 show would get 10% of the booking, and because of that, would also get an additional 15% as a management commission. Some of the major music-related unions franchise and regulate major booking agencies and require this separation of duties. So the inclusion of a statement in the contract between the artist and manager that says it is not a responsibility of the manager to procure employment for the artist is intended to give the manager something to point to if he or she is accused of violating laws or regulations relating to booking performances.

As a practical matter, many agents will not book small venues because it isn't worth their time to trade calls with club owners, negotiate and issue contracts, and collect a small amount of money for the time they need to invest to secure the booking. In these circumstances, the manager seeks a small local agency to handle bookings, or sets up a small subsidiary firm for the specific purpose of this kind of booking. Bookings at larger venues in most cities are available only through a well-connected agent.

A final point about the California Talent Agency Act is that the law specifically permits the manager of an artist that performs music to negotiate a recording contract on behalf of their client. This clears up any implication that a record contract is "procuring employment," and it preserves one of the most important responsibilities of an aggressive and effective artist manager.

EXCLUSIVITY

An artist usually agrees in the contract to have only one manager, and it is exclusively the manager whose signature is on the contract. On the other hand, most managers require that their services be on a nonexclusive basis, meaning that they can develop their own roster of clients beyond the artist who is the subject of the contract under consideration.

Any artist manager or management company should have a range of clients whose careers are in different stages of development. Many active recording careers end within a very few years and become primarily performance-based. So the savvy manager has new artists and midlevel artists who are developing successful careers to replace those whose careers have matured. And for this reason, a manager should be on a nonexclusive basis with new artists that they sign. An exception to this is a major act that requires full-time management by a single manager or management company.

POWER OF ATTORNEY

Power of attorney is a document that gives someone the legal authority to sign documents and make contractual commitments on behalf of someone else. Most powers of attorney are very specific about which contractual agreements that the manager may sign for (obligate) the artist, and there may be limitations set that define the monetary amount within which a manager may obligate the artist. For the artist, the power of attorney may permit the manager to:

- Sign contracts for performances and appearances on television and in advertising and motion pictures

- Endorse checks, deposit money, and pay employees
- Hire and fire the artist's team members
- Approve the use of the artist's likeness or voice for advertising, merchandise, and ringtones

New artists may be reluctant to give a manager such broad authority, and many artists prefer to separate the business management function from the general management responsibilities.

THE MANAGER'S PAYMENT FOR SERVICES

Artist managers are most often paid by a commission that is based on the gross earnings of the artist. The amount of the commission paid by an artist to a manager is typically 15% but can be as high as 25% for new artists and as low as 10% for established artists. New artists do not have much earning power at the beginning of their careers, so managers are paid a higher commission rate in order to make their investment of time somewhat worthwhile. Even at the higher rate, managers of new artists will have little or no earnings at first. Managers often forego commissions to which they are entitled, but keep ledger entries anticipating a time in the future when artist earnings will be available to pay them. Managers of established artists may be comfortable earning 10% because the smaller percentage is applied against a much larger income stream and therefore can generate substantial commission payments for their services to the artist.

A variation of the straight commission on gross earnings is one in which the manager may be paid a commission based on net earnings of an artist (meaning the gross earnings of the artist after expenses). It can, however, become difficult to determine what those "net" earnings are, and the manager will want to have some control over expenses that are deducted from gross earnings. In another variation of commission payments, some groups may seek to cap the earnings of a manager because they may be paid more than any single member of a group without limits being placed on what the manager can earn from performances by the artist.

In this section of the contract, the artist and manager also determine which of the artist's earnings are to be commissionable by the manager. For most new artists, all earnings are commissionable from the beginning. The most cited exclusion from earnings is publishing income resulting from songwriting, but the manager who becomes responsible for negotiating a publishing contract for the artist will generally be entitled to commissioned earnings from that agreement. Advances to the artist that are not included as part of the cost of creating the recording are immediately commissionable by the manager because they are viewed as prepaid royalties to the artist, but tour support money is not permitted by the record company to be commissioned by the manager. In the latter instance, the company feels that the advance is provided to offset the costs of touring rather than to provide an income stream to the artist.

Managers may agree to be part-time career directors for artists, and some manage by charging a monthly retainer to the artist. These arrangements can be less costly to the artist, but the additional "price" the artist pays is a part-time commitment to their career rather than a full-time involvement in their career success.

When discussing this section of the artist management contract, it is helpful for the new artist to understand how compensation flows to the manager and how payments to artist team members reduce their earnings. It can be a surprise to artists to learn how much is taken from their earnings unless they have been prepared by a candid discussion. We discuss more about this in later chapters, but here is an example: before other expenses are paid, a $10,000 booking requires a 10% payment to the agent and a 15% payment to the manager. The remaining $7,500 must pay the artist and all performance expenses. It may seem excessive to the new artist, so this becomes a genuine reality check for him or her. It is better to have the discussion at this point, rather than at a time when both are trying to build a relationship based on candor and mutual trust.

EARNINGS FOLLOWING THE CONTRACT PERIOD

Artist management contracts should, for the exclusive benefit of the manager, have what is termed alternatively a *sunset clause* or a *blackout period*. It can best be described as a severance package for the manager. Regardless of what this section is titled, it creates a time after the contract ends during which the manager will earn commission for work that is already planned and is under contract. For example, if a three-year management contract ends and is not renewed and the artist has forty show dates booked for the next year, the sunset clause allows commission to flow to the manager for earnings generated by those performances. These postcontract provisions may be structured to step the earning rate down to perhaps 10% for the first six months, 5% for the last six months, and ends after a year. The artist and the manager may also include special provisions on how royalties from songwriting, recording, and special licensing will be handled when the contract ends. These terms are negotiable between the artist and the manager at the time the contract is negotiated, and both the artist and the manager should take special care to foresee as many special circumstances that may affect postcontract earnings for the manager.

Another contract provision within the sunset clause that the manager will want is the circumstance in which the artist manager will benefit from deals he or she has initiated on behalf of the artist that are not under contract when the management contract ends. For example, if the manager has been developing a major sponsorship deal that is not completed until a year or two after the management contract ends, he or she will be able to claim a commission for any prior work in developing the deal.

The need for a sunset clause in the contract becomes especially important when the artist and the manager both agree to end the contract prior to its expected termination.

When this happens, there is pressure on the artist to find someone to continue to manage their career, and having a workable sunset clause minimizes the impact of paying commissions to both the former and the new manager at the same time.

The sunset clause is also important in the event of the death of the artist because it builds in an income stream for the manager and gives the artist's estate guidance on the wishes of the artist. When Elvis Presley died in 1977, his estate planned to end the relationship with manager, Tom Parker, in part because of the 50% commission he was taking from the artist's earnings. However, the artist management sunset clause specified that Parker would continue to receive the commission even after the artist's death. The estate sued Parker, and four years later his commissions ended. The two sides eventually settled out of court in 1982 (AP, 1982).

THE MANAGER'S EXPENSES

Many relationships are strained because of money issues, which is why contract sections dealing with expenses and compensation for services are very important. The artist and manager should be candid and very clear in the contract about how much money will be paid to the manager, under what circumstances, and when it will be paid.

A section of the contract describes in detail how the expenses of the artist manager are to be handled. *Expenses* are out-of-pocket expenditures made by the manager specifically on behalf of a particular artist. Expenses usually do not include office rent for the manager or other costs that are paid by a company that are considered the overhead needed to operate a business. For example, a manager who is required to fly from New York to London to meet with a group to discuss a European tour for the artist will seek reimbursement from the artist for all monies spent from the manager's personal funds to cover the cost of this trip. Expenses for something like this could become considerable, so the artist may place within this section of the contract a limitation on the amount of expenses beyond which the approval of the artist is required before the expenditure is made.

An example of what happens when limits are not placed on expenses involves Sir Elton John and his former manager, John Reid. Following his world tour, Sir Elton sued his management company and others in 2000 for what he deemed excessive charges for their personal expenses during the tour. His settlement with Reid required the former manager to repay five million dollars in excessive expense claims that had already been paid to him (BBC, 2000).

OTHER SECTIONS

The contract should include provisions for the artist and the manager to agree to insure each other's lives if they wish. Because of the important nature of their business relationship, both may be viewed as having an insurable interest in each other's lives and should have the expressed permission to purchase at their own expense a life insurance policy on each other.

A section of the contract includes a provision saying that the artist will not permit the manager to assign the responsibilities of the agreement to anyone who works in the management company or to anyone who might acquire the assets of the company. Sometimes referred to as a *key man* clause, this assures the artist that there will be continuity in their management, and that any changes within the management company that would deny the artist the services of their manager will be a cause to terminate the contract. This provision permits the manager to delegate authority to subordinates such as a tour manager but does not permit another manager to take responsibility for the artist's career management.

Many small management firms allow themselves to be acquired by a larger company, and this section of the contract permits the transfer. What is not permitted by most management contracts is the right of the manager to assign, or sell, the management contract to another manager. As mentioned earlier, the relationship of artist and manager is unique and personal, and an artist will not permit that relationship to be sold.

No relationship is always perfect; there will be occasional disagreements. A section of the contract should give guidance on how the artist and manager will handle these times. Requiring arbitration is a way to remedy disputes that can limit legal costs and often keep disagreements from entering the public arena associated with court systems.

If the manager and artist want to modify the contract, they should put it in writing. The contract should include a section that requires any changes to the agreement to be made in writing. Changes in writing include the advice of the individual attorneys and minimize any misunderstanding of the consequences of the change.

If the artist is made up of a group of performers, special language is necessary to protect the rights of the manager in the event that a member quits or is replaced by the group. The manager should have the option to end the management relationship and cancel the contract within 30 days of any changes in personnel. The contract should also require the member who leaves the group to be managed by the manager in the event they decide to pursue a solo career. This section is intended to protect the manager from being legally obligated to manage an ineffective artist yet protect the investment made in developing the career of the member who left to pursue a solo career.

CONTRACTING WITH A MINOR

One of the realities of the American legal system is that people who are under 18 years old cannot be held responsible or liable for contracts they sign. Until recently, that meant that a minor artist—someone under age 18—could sign a management or recording contract and decide to end it. Terminating a contract would have no consequences to the minor, and the manager, publisher, concert promoter, or record label would have to absorb any losses that resulted.

California and Tennessee were the first to create laws designed to make companies less reluctant to enter into business arrangements with artists who are minors, and at the same time protect some of the earnings of the artists until they turn 18. Tennessee's law is called the "Tennessee Protection of Minor Performers Act," and comes from the Tennessee Code Annotated books under Title 50. The law permits a minor artist to contract with an individual or company and be bound to the agreement. The law, however, requires the contract to be approved by a judge, who will then regularly monitor the effects of the agreement on the career of the minor. In this way, a third party is involved to assure that the minor is protected, yet there is also a measure of protection for the manager or company. Tennessee law requires that contracts created under this law include a provision that holds a minimum of 15% of the earnings of the minor in a trust account until he or she turns 18. California's law works similarly, but the trust withholding requirement is 25%.

New York has a law similar to the two mentioned, but it places a limit on the length of most contracts with minors to no more than three years from the date the contract was approved by a court (Berry, 2005).

A CONTRACT EXAMPLE

A copy of a form that is used for contemporary artist manager contracts is included as Appendix B to this book. The form includes the key components of a management contract but should not be taken to represent the final contract that an artist and a manager should sign. What is shown in the appendix is the shell of many of the negotiating points that should be included, depending on the nature of the agreement the manager and the artist seek to memorialize with the contract.

The bottom line is that the artist and manager should employ entertainment lawyers to design their contract.

References

Associated Press. (1982). Presley Estate and Manager Reach Accord on Payments. *New York Times* Retrieved from http://www.nytimes.com/1982/11/17/us/presley-estate-and-manager-reach-accord-on-payments.html.

BBC. (2000, November 15). Retrieved from http://news.bbc.co.uk/2/hi/uk_news/1024745.stm.

Berry, L. (2005). http://www.berryentertainmentlaw.com/articles/contract_minors.pdf.

SoundScan. Retrieved September 20, 2010.

A primer for the artist manager

6

An effective manager working as an advocate for an artist requires an understanding of the general elements of planning and an appreciation of the value of plans in guiding an artist's career. This chapter is an introduction to planning and includes specific guidance on creating and budgeting for a plan.

A weakness found in the music business is the lack of any formal planning for artists, especially new artists. Veterans of the music business—both artists and their managers—have a short-term and a long-term vision for managing careers, which are frequently not formalized in any written form. Artist manager Stuart Dill says, "Developing careers and plans, and actually putting them on paper and in writing I think is always a good idea. I don't think it is always the norm. I think you'll be surprised at how the industry has not had a formal approach as you would [if you were] a Fortune 500 company" (Dill, 2004). Even today, many in the music business agree with Dill's assessment. For the new artist who is ready for the commercial exploits of his or her music and is signed with management, a formal written plan is especially important. Chapter 12 of this book guides artist managers through the creation of such a plan. This chapter begins with goal setting.

SETTING AND ACHIEVING GOALS

Perhaps the most often spoken goals by people everywhere are New Year's resolutions. We heard the resolutions in early January—and perhaps made some ourselves—knowing full well that there wasn't a chance that they'd all be kept.

In part, this is because they were terribly ambitious, and in part because they were made without a plan or timetable to make them achievable. As the next year approaches, the resolution remains unmet and as the sound of *Auld Lang Syne* rings in the New Year, the same goals are set for yet another year. Promises you make to yourself toward achievement of goals without a plan and a timetable suggest that they are more like dreams rather than intended results.

Managing careers of artists converts their dreams of success in the music business into a reasonable reality. But without a plan and accompanying timetable, coupled with a shared vision of the manager and the artist, an artist's career becomes a dream that blindly searches for opportunity. It is important that the goals of a plan and the way to reach those goals are the result of collaboration between the two. The manager should push the artist, as the U.S. Army says, "to be all you can be," and the artist must view the plan as being one that is reasonable and achievable. Without the mutual agreement on the plan, the artist may not take genuine ownership in the viability of the plan and will see it as the manager's vision, not theirs together.

An often-quoted line from Michelangelo fits well into to our discussion of setting goals. "The greatest danger for most of us is not that our aim is too high and we miss it, but that it is too low and we reach it" (Chandler, 2005, 51). There is always the temptation to take the safe road and set goals that assure success. However, doing so sets the stage for underperformance or underachievement based on the talents of the artist. The manager must negotiate goals into the plan that the artist genuinely feels are attainable with the guidance of the manager, and they must be a true reflection of the potential of the artist's talents.

Defining a set of goals by and for the artist is the starting point for creating a plan. Goals chosen by the artist might be securing a recording contract, learning to play piano during stage performances, learning to write songs that have commercial appeal, developing a commanding stage presence, or acting in movies. But as these goals are set, they need to be very specific. For example, stating a goal as general as "acting in movies" doesn't have a beginning or an end, which makes it more of an idea than a goal. A goal for the artist should be stated as specifically in the plan as this: the artist will begin receiving coaching in acting at the end of next year's tour with the goal of securing a part in a motion picture 12 months later. Generating specific goals takes out the "pipe dream" aspect of goal setting and creates accountability for both the artist and the manager.

Planners use a common metaphor that says a goal is the target you are trying to hit. It is the result you are trying to achieve. It is the goal—the target of your energies. In order to achieve a goal, you need a set of strategies. Strategies are the plans you make to reach the goal. One of the best ways to develop a good set of strategies is to create a timeline beginning with the goal and work backward, determining what must be done to reach the goal. Strategies are lists of general steps that the artist and manager must take in order to get the desired result found in the goal. An example of a strategy is to create and practice a stage show with the goal of being booked to open for a headliner. Ken Kragen has a chapter titled "Backward

Thinking for Forward Motion" in his book, *Life Is A Contact Sport* (1994), which offers an excellent guide for developing effective strategies to reach goals. Among his suggestions is to motivate gatekeepers in a career path to say "yes." The music business is one of those high-reward industries that has many gate-keepers, and it requires the manager to plan for how to identify and deal with them along the way.

The final piece to the planning model is the use of tactics. These are the things the artist and the manager do each day to implement the strategies in order to achieve the goals. With the goals set and the strategies developed, the manager's personal planning tool—a smart phone, a day planner, a calendar, a computer, or an old-fashioned notebook—will have daily entries focused on executing the strategies. It will note telephone calls to be made, emails to be sent, meetings to be held, and a follow-up step for everything.

Chapter 12 presents software that can be used by the manager to create charted timelines that visually display the goals and related strategies. These applications are effective tools to help the manager manage time quickly, and to show the artist the commitment they have both made toward the artist's success.

PLANNING A PERSONAL BUDGET FOR THE ARTIST

The manager who does not create a reasonable personal budget for the artist, especially a new artist, will be spending valuable time trying to patch up relationships with bankers, credit card companies, and others to whom the artist owes money. Sitting down with the artist and developing a monthly budget will show the artist spending limits as well as give the artist guidance on being financially solvent. It is difficult for artists to be creative in their art if creditors are applying pressure and making demands for payment. Keeping them financially solvent eliminates a possible distraction and helps keep them focused on their art.

The outcome the manager wants is for the artist to understand the basic idea of personal budgeting: monthly Income = monthly Expenses. In a perfect world, the artist will have more income than necessary to cover expenses and will be able to put some into savings. But the simple model noted here is the minimum outcome of a well-planned budget based on available income and necessary expenses.

A budget, like any other plan the manager creates, must be done in collaboration with the artist. The artist must see the need for it and must agree to the limits of the plan. A lot of personal budget templates are available that make excellent tools for assembling a personal budget. An especially useful one is presented by the Consumer Credit Counseling Service of San Francisco (CCCS) and is available on their website at *http://www.cccssf.org*. The budget can be arranged by the week or the month, depending on which works best for the artist. Monthly budgets are easy to prepare, but they begin to fall apart by the middle of the month when the artist learns that there isn't much money left for the remaining two weeks of the budget period. Initially, then, a weekly budget will keep a shorter-term view of where money is being spent, and it gives the artist a way to develop the control necessary to be responsible with prescribed spending limits.

The manager and artist should plan for the following areas as major areas of the budget:

- Savings
- Groceries
- Restaurants
- Laundry/Dry Cleaning
- Medical/Dental
- Automobile/Gas/Parking
- Other Transportation
- Child Care
- Personal Care
- Clothing
- Postage/Bank Fees
- Entertainment
- Books/Music/Video
- Cigarettes/Alcohol
- Gifts/Cards
- Home/Garden
- Church/Charity Contributions
- Other[1]

The guidelines by the CCCS for a budget by general categories are:

- 35% for housing
- 15% for debt
- 10% for contributions
- 15% for transportation
- 25% for all other expenses

It is important for artists to understand that a budget is based on their net income, meaning that money for their personal budget is available only after taxes have been taken out. The nature of the work of many artists is such that they are paid the full amount of their earnings when they are earned, and taxes will not have been deducted from the amount the artist receives. What will happen in January is that a U.S. Internal Revenue Service (IRS) Form 1099 will be sent to both the artist and to the IRS by the various employers, which is a reminder that the earnings were paid in cash or by check. No taxes were withheld from these amounts paid to the artist. The artist will be required to have funds available to pay any income taxes that are due from amounts reported on all of the 1099 forms they receive. David Darnell Brown performing as rapper Young Buck owed the IRS

---

[1]Consumer Credit Counseling Service of San Francisco (2010).

$164,000 in unpaid back taxes in 2010, so his home was raided and some of his personal property confiscated, which was set to be auctioned to pay the tax debt (Gee, 2010). An artist manager must help resolve situations like this or, ideally, make sure they never happen at all.

In addition to the Federal taxes payable in the United States, the artist must be prepared to pay any state income taxes for amounts earned from performances in each state he or she worked in during the previous calendar year. It is important that the artist understands that a budget category will be created to hold back anticipated taxes that will be due.

As the artist begins to live with a budget, the manager and the artist may discover that adjustments need to be made to ensure it is an accurate reflection of the artist's requirements.

PLANNING AND BUDGETING AN EVENT

Throughout one's career as an artist manager, there will be countless planned events, both large and small. Regardless of the size of the event, it requires great attention to detail and a realistic budget because it's an important component of promoting the career of an artist. The range of events can be from a small "meet and greet" with a tour sponsor involving the artist and a handful of other people to something as large as a performance before 40,000 people. Although the latter example is often planned by a promoter and an agent with final approval by the manager, supervising performances and other appearances is the responsibility of the manager, and the manager must know what elements must be planned into them to assure their success for the artist.

For several reasons, a plan for any event should be written. First, a manager who has several artists will have a number of events in the planning stage, and a written plan will keep them organized and separate from each other. During those frequent hectic times, keeping the plans separate and in writing will minimize the confusion for the manager and those who are assisting with them. Having a plan in writing is especially important when the manager must share information with others. A range of distractions from illness to other unexpected emergencies becomes easier to manage if a clear and complete written plan can be assigned to someone else to handle. The written plan should include a lot of detail:

- The time, date, and place
- Purpose of the event with the expected outcome
- All contact information for the manager, including email, landline telephone number(s), home phone number, cell phone number, instant messenger address, fax number, and any other wireless access address
- Similar contact information for the manager's primary assistant
- All special requirements for the event (food, beverage, equipment, personnel, room setup)

- Detailed contact information for all service and equipment providers for the event, including after-hours and emergency telephone numbers
- Contact information for all backup service and equipment providers
- Full transportation and travel information, even if it is for a local event
- A budget detailing expected expenses and who will pay for each element of it

Successful events require considerable attention to detail. Each element of a planned event involves calls, follow-up calls, confirmation, and reconfirmation as the date approaches. The event plan we discuss on the following pages, the responsibility of something as seemingly routine as issuing invitations, is highlighted to demonstrate the amount of planning and coordination necessary to simply invite people to a showcase.

The cities of major music centers have venues that routinely showcase artists for the purposes of securing management, a recording deal, or a publishing contract. The showcases can be one of the regular performances by an artist at a venue at which special guests are invited to determine whether they'd like to have a business interest in the performer. Sometimes they are as simple as an audition in a conference room. Other times they might be held at a performance rehearsal facility, and sometimes they are invitation-only showcases presented at small clubs. The event plan example in this section of the book is for a showcase for an artist who is seeking interest by a large independent label or by a major label for a recording contract. Because the objective is to enter into a business arrangement, one of the key music business centers—Los Angeles, Nashville, New York, or London—is the likely location for a showcase. And regardless of where the event is held, the planning for it has the same considerations. Depending on the available budget, elements of the plan described in the following section will require modification or elimination. A showcase at a club that is part of a regular engagement for the artist has little additional cost; a showcase planned as an exclusive event for the artist can involve considerable expense. Each can be effective, although the exclusive event focuses on the artist, and is designed to be purely an industry event rather than mixed with club patrons.

AN EVENT PLAN
When and where

Before an event can be scheduled by the manager, time must be blocked from the artist's schedule to prepare for it. A showcase to seek a label deal will require at least six to eight weeks of preparation from the date the decision is made to have the event until it happens. Before the date is confirmed, the manager must look at the general calendar for the music industry to be sure that the key people to whom invitations will be sent aren't committed to something else. If, for example, there is a major award show scheduled or an industry conference set for the artist's musical genre, the showcase should be planned at another time. Conflicting events, especially locally, can kill the intended result of a showcase. Competing for the attention of gatekeepers at industry conventions is expensive and often unproductive because of the number of people present and the amount of talent present.

Specific timing ideally includes a late afternoon event that closes out the business day for invitees. If you offer refreshments to the guests, a broader range of beverages can be offered at the end of the day.

The most efficient showcase plan is one that requires about an hour of the invitees' time. Thirty minutes is set aside for people to arrive, have light refreshments, and network with peers. The last half hour will include a twenty-minute performance by the artist with ten minutes reserved to navigate the room with guidance by the manager to connect with key people who attend. A showcase planned at 5:30 p.m. gives invitees a half hour to network with peers, hear the showcase, and be on their way home or to other evening activities by 6:30 p.m. It is an efficient use of their time and a convenient way to end the business day.

The venue the manager chooses should be one that is large enough for the event, but small enough so that it is easy to give the perception that there is a big attendance. An ideal venue will be conveniently located near those who are invited, have adequate parking, offer the use of the venue at no charge for a guaranteed food and beverage purchase for the event, and seat 100–150 people in front of a permanent stage. The venue does not need to be a hangout for music business types, but it should be among those defined as being "hip" or "cool," such as the House of Blues in Los Angeles or New York, 12th and Porter in Nashville, or The Borderline in London.

Invitations

With the time and place set, it is now possible to create and issue invitations. The surest way to a successful showcase for the artist is to use a combination of the old and tried ways with the inclusion of technology. Three weeks before the event, mail the invitations using the traditional postal service so they will arrive two and a half weeks before the event. The following week, send a follow-up email that includes an image of the invitation. To those who have not responded to the first invitation and first email, send a reminder email the week before the event reminding the invitee again that the showcase is coming up.

Every part of a successful plan has countless steps, and the effective manager anticipates each element of each step. For example, as easy as this seems, issuing invitations has a considerable number of steps that will require attention to detail:

1. What format will the invitation take? Will it be an invitation inside an envelope or will it be a postal card? If it is a postal card, what size will it be? Will it be printed in four colors (that is, full color)?
2. What will the invitation say? Will images be required for it and from where will you get them? Who will proof the text?

3. How will invitees confirm that they will attend? Should they call the manager's office or will an RSVP telephone line be set up to take reservations? Who will compile the RSVP report on a daily basis for the manager's review?
4. Who will print the invitations, someone in-house at the office or a printing company? How many bids will you take for the printing service? What is the printer's lead time to be sure the invitations are ready on time? Who is your back-up printer in the event the invitations are not ready on time?
5. Who will receive the invitations? How many invitations must be issued to ensure the size and quality of a crowd at the event? Will the artist manager's list include everyone who should be invited or should the manager also borrow a list from a major publicist to assure adequate coverage of the industry? Who will eliminate duplicate names? Does the manager have software and expertise to print labels from the database?
6. Who will assemble the invitations and prepare them to post? How much postage must be purchased and who will purchase it?
7. Who will create the email follow-ups? Is there a complete email list of invitees? Does the manager have the software and expertise to send an email blast that can penetrate corporate firewalls to get to recipients?

Although this list demonstrates the kinds of tasks and detail necessary for just the invitations, it does not include the telephone calls, voice mail, and emails necessary to complete each of these steps.

Food and beverage

Often referred to as *hospitality* in the meeting planning industry, the kind of food and beverages served at an event like this depends on the budget available. Venues often supply the event space at little or no cost, provided that the manager purchases snacks and beverages from them rather than having them catered. An open bar for a one-hour event will leave the artist manager exposed to the consumption of the crowd. If beverages cost an average of $4 each, a crowd of one hundred can easily accrue a bar tab of US$1,000 in an hour plus an additional $200 tip to the barkeep. On a larger scale, the author held a major music industry showcase and dinner for 1,600 people that included an open bar for an hour. The bar tab for that 60 minutes was $27,000, although only $15,000 had been budgeted for it. At the settlement with the venue, the amount was negotiated down, but thereafter beverage tickets were provided for similar events as a way to control costs. The use of beverage tickets is encouraged for any event at which the food and beverage budget must be closely watched.

If food is planned for a small showcase, it must be within the limits of the budget. Depending on the city in which the showcase is held, catered snacks and light finger foods can cost between $10–$20 per person, including tax and tip. For a small showcase, however, food is not necessary if the budget does not permit.

The performance

The manager should assemble a list of the requirements for the performance, many of which depend on the venue. Some performance venues offer only a stage with the expectation that the performer will bring in the necessary sound and lighting equipment. Other venues have "house" sound and lighting systems with technicians to operate them for the artist, and these accommodations may include a small charge or be offered at no cost.

If the stage is not adequate for the performance, the manager must rent appropriate staging and have it erected for the performance. Some venues have no stage, and the performance is made from floor level. Some have minimal staging that is more like a riser than an actual stage. Others have staging that puts the performer above the audience, permitting him or her to be seen and heard very well. Besides the elevation of the stage, the manager must be sure that there is adequate space for the stage plot that is planned for the event. (A stage plot diagrams the location of each performer and their associated equipment.) If more space is needed, the manager must order it.

If the performance is by a group that includes its own band members, it may be unnecessary to hire additional players for the showcase. However, if the artist does not have players who regularly perform together, a budget allowance is necessary to pay the players for rehearsal and performance time. Costs that may be incurred with the performance could also include cartage (paying someone to transport instruments and other equipment), instrument rental, and technicians to manage the lights and sound.

Promotion

Those who attend the event should be provided with promotional materials, often in the form of a press kit, to carry away with them after the event. It should be given to all invitees as they leave the event. Encourage them to take two if they need them, keeping in mind that many who attend showcases are surrogates for decision makers, and they will be sharing the experience with other staff at their company the next day.

In a brief showcase, the name of the artist is typically mentioned at the beginning and at the end of the performance. Because they are relatively new to the industry, it helps to keep a banner with the name and website address of the artist on stage in front of the audience before, during, and after the performance. Heavy-duty banners are relatively inexpensive and can be used for subsequent performances.

The final piece of promotion is often the most expensive but can also be the key to a successful event: hiring a good publicist. Entertainment publicists can promote the event to the industry and can manage the event for the artist manager. Publicists will be able to help create a list of key invitees, provide addresses, handle RSVPs, make follow-up calls to those invited, and manage the event on the premises the day of the event. They know who the key people are in the industry and often have the relationships with them that are necessary to get them and/or their staff to the showcase.

A SAMPLE BUDGET

The most economical way to showcase talent is to make it a performance at a venue at which the artist is already booked. Guests who are invited can be asked to present an invitation at the door, or a list of invitees can be provided to the person at the door of the venue. Accommodations for invitees can include reserved tables with directives to the wait staff to give complimentary beverages to them (with the bill coming to the manager).

An exclusive industry showcase is the most expensive way to showcase, and is most appropriate for the aspiring artist who has access to enough funding for an event of this type. Typical costs an artist could expect to pay for an event that plays host to 100 industry people are shown in the following table:

Budget Item	Cost (U.S.)
Beverages	$1,200
Food	$1,000
Rented sound and lights with technicians	$3,750
Invitations (designed and printed commercially, postage)	$850
3′ × 5′ hand-painted vinyl banner	$150
Musicians (5)	$1,200
200 press kits	$500
Publicist	$3,500
Cartage, instrument rental, miscellaneous	$450
Total	$12,600

The $12,600 for this exclusive artist showcase allows you to create an event specifically targeted to gatekeepers of the artist's career. Events like these can generate industry interest, and depending on the talent of the artist and the skill of management, they can be productive.

The question most often asked about a showcase event—whether it is an exclusive artist showcase or a regular performance to which key industry people are invited—is, "How do I pay for it?" Artists have several options to find funding sources. They may have savings that are put aside for their career development; they may have friends or family members who are able to assist in paying the costs; they may have a sponsor who regularly associates with their performances and is willing to underwrite an event like this; and the artists may be able to take out a loan to cover the showcase costs. Some artists have "super fans" who are investors or "angels" and are willing to participate in the financing of the artists' careers in exchange for an earnings return from their investment when that career takes off. Every artist that has a well-developed fan base also has a few sincere fans who will say, "Please let me know if there is anything I can do to help you succeed," and these people are ones who can help cover showcasing costs. As noted earlier in this book, the artist manager should ask the artist to remember and note anyone who falls into this category, and the manager should then follow up and ask for support.

Planning for any event, whether it is an artist showcase or a meet-and-greet preceding a performance, requires close attention to the budget. If the plan results in a budget that the artist cannot afford, then the manager must decide whether to modify the event so that it becomes affordable or consider not holding the event at all. Remember that budgeting is a matter of setting priorities with available funding, and above all, a plan must be affordable.

PLANNING TOOLS

There are a number of organizing and planning tools used by artist managers. They include paper planners from Day-Timer® and Franklin Covey®. This style of planner has accessories and sophisticated tools that help the artist manager stay current and plan for the artist as well as for all other company clients. This style of planner sells annual updates with printed pages, binders, and useful reference pages.

Another planning tool is Microsoft Outlook®. It provides many of the same functions as paper planners, but it is software used on computers, smartphones, PDAs, and other handheld devices. Outlook can be an effective information and communication management tool and works as an adequate planner for the artist manager.

People who are successful in the sales and promotion professions will say that they love the business they're in, but they hate the paperwork. However, planning and its associated "paperwork" can give artist managers the competitive edge at being an effective advocate for everyone on their artist rosters.

References

Chandler, S. (2005). *Ten Commitments to Your Success*. Barndon, OR: Robert Reed Publishers.

Consumer Credit Counseling Service of San Francisco. (2010). http://www.cccssf.org.

Dill, S. (2004). *CMA's Music Business 101*. Unpublished video.

Gee, B. (2010, August 31). Young Buck Raid Netted Thousands. *The Tennessean*, p. 2B.

The artist as a business

7

Any business requires an understanding of what product or service it is providing and to whom it is providing it. It is a basic mission statement. With the mission established, it becomes the job of the manager to refine the definition of who the expected customers of the artist will be, to employ the people and other resources to create a support team, and to develop a plan to create a relationship between the artist and their fans that includes commercial activity.

It is clear that among the most important functions of the artist manager is being the promoter and advocate for the artist and their talents. This is another way of saying that the artist manager is responsible for developing the artist as a brand, marketing the artist, and exploiting their talents—which is why this and many of the following chapters include sections on marketing and promoting the artist as a commercial entity. This chapter begins with a look at target marketing for the artist manager followed by the creation of the artist's support team.

UNDERSTANDING TARGET MARKETS

The job of artist management is to direct the career of an artist, and a very common activity of the manager is promoting and selling the musical talents of the artist to people willing to buy music and tickets to performances—this means that there is a lot of marketing going on. The manager must keep a constant understanding of what is driving the buying decisions of music consumers, especially within the genre of the artist, and must translate that information into revisions to the artist's career plan. The manager should also be on the watch for opportunities to redefine the artist's target market in order to build a larger fan base.

DEFINING AN ARTIST'S TARGET MARKET

A *target market* is made up of consumers and potential consumers with whom the artist's music connects creatively and commercially, and they are the fans who have the willingness and means to buy music, tickets, merchandise, and related ancillary products from the artist. This target market makes up a segment of the larger general market of music consumers. In the marketing profession, targeting this smaller segment of a broader market is defined as *market segmentation*.

WAYS TO VIEW MARKET SEGMENTS

There are numerous ways to look at segments of a target market. The savvy manager will not assume that the marketing department of a record label is viewing the full career of an artist in terms of the label's target market segment. As we know, the traditional function of a label is marketing and distribution with the goal of selling recordings—not necessarily selling tickets and merchandise (although 360 recording contracts have changed this). So it is important that artist managers keep a continuous look at trends in the genre, in music generally, at new product technology, and in pop culture to find all opportunities for their artists that might be in the marketplace. The manager should then adopt those opportunities into the formal career plan of the artist to be sure they have become integrated into the goals set by the manager and artist.

A good definition of market segmentation is "the process of dividing a large market into smaller segments of consumers that are similar in characteristics, behavior, wants or needs" (Hutchison *et al.*, 2009, p. 20):

- The broadest way to view a market segment is to define it by demographics such as age, sex, race, religion, and other similar criteria. This is the segmentation type used most often because it is relatively easy and inexpensive to create a target definition and it can help create distinctive target groups, although it can be somewhat shallow when compared with other segmentation methods.
- Geographic segmentation is another standard method of defining consumer groups, but it doesn't reveal as much about potential customer groups as other ways.
- Psychographics segmentation is viewing a market segment based on lifestyle characteristics of the buyers of music and tickets.

- Behavioristic segmentation looks at why consumers engage with a product, how they use the product, and what creates their loyalty to the product.

The last two methods of segmentation that help define the target market for an artist require considerably more thought and ongoing research by the artist manager in order to be useful. However, artist managers who stay current in their understanding of generational attributes of the target market in areas of psychographics and behavioristic segmentation will be able to find subtle ways to reach consumers that others will not. An extension of this is to be continuously aware of changes within peer groups that might establish new attitudes to find opportunities, or to head off a consumer group's flight from favoring an artist. There is no question that psychographic and behavioristic segmentation approaches can be expensive to include in career strategy development for the artist; however, where resources are available, they can offer a distinct advantage over artists who don't use these sophisticated tools to more closely define their target market.

An example of the importance of target marketing is the recording success history of John Mayer. In 2001 his first album was released, and it performed extremely well in the marketplace. His three subsequent albums were successful projects, but each produced much lower sales than the previous album. The possible causes for the decline in sales are numerous, but it raises these questions about the target market. First, has the marketing approach been modified to acknowledge that the 22-year-old fan who bought his first album is now over 30? They're in a different place in their careers and they are different consumers now. And second, what has been done to embrace the special target market consumer characteristics of those who are now 18–22 years old?

An informed understanding of the target market segment can be one of the most productive and efficient tools an artist manager can use in managing the career of an artist.

BRANDING AND IMAGE

A *brand* is typically the name of a product that consumers identify with in terms of the benefits it provides, so the artist's name and everything associated with it create a brand name in the minds of the target market/fans with the benefits of being entertained and associated with someone many admire. The name of the artist is a distinct brand, and like all brands it becomes the sum of all of the experiences the fan has had with the artist's music. Artists who present themselves as a brand including the associated image distinguish themselves from others and become more recognizable through their distinct approach to music and performances. With the large array of cable music channels, video websites, and smartphone apps like Vevo and YouTube, the artist has become multidimensional and has taken on the qualities of the purest form of a brand. Although brand identification with labels such as Chess, Motown, Atlantic, and Def Jam has created a certain expectation by the consumer over the years, the brand relationship the music and ticket buyer has now is with the artist, not with the label.

A key component of branding is image. A brand image is the way people feel about a product, and in this case, the artist. It is their emotional attachment to the artist based on their music and performances and how people respond to them. The image of the artist should be a reflection of their own personal values, but then it should also be a reflection of the values of those who buy the music and concert tickets—the target market.

We live in a world that requires us to filter tens of thousands of commercial messages every week to quickly determine whether they are relevant to us and our lifestyle (CNN, 2007). With such a continuous assault on our attention, we've become very adept at spotting phony communications that are pointed at the masses with little originality. It's the same with today's commercial music. As a brand, artists must be unique but fitting for the genre, consistent in the quality of their work, and genuine about their artistry. So it becomes the job of the artist manager to help artists refine their image so it has a commercial edge but still maintains the values the artists bring into their work.

The symbol of the brand comes in the form of a trademark or service mark or both. A *trademark* is a symbol that represents the brand and distinguishes it from other products. Artists who choose a symbol as their trademark should also trademark the text of their name in order to protect it in its commercial use. Where trademarks define the sources of products, a service mark gives the source of services, and in the case of artists, the service they provide is entertainment. Both kinds of marks generally are protected if they are regularly used in business, but for assured protection of their exclusive use by the artist they should be registered with the U.S. Patent and Trademark Office.

Related to trademarks is a doctrine referred to as an individual's right of publicity. This generally means that people have the right to control the commercial exploitation of their names, their image or likeness, or some other distinguishing aspect of their person even having a very distinguishable singing voice. In 1989, Bette Midler declined an offer to appear in an automobile commercial, so the advertising agency hired a sound-alike singer instead. Midler sued and won $400,000 with the court saying that she had a "property right to her distinctive singing voice" (*US News & World Report*, 1989).

Many states, including California, New York, and Tennessee, have passed laws or accept the common-law practice of protecting a person's right of publicity. This protects the exploitation of an artist's brand or image for commercial purposes without permission to do so. In the UK, the principles behind the right of publicity have not been supported, although the courts are more willing to consider the concept today and research on the right to publicity continues at the University of Edinburgh School of Law (Black, 2010).

A final note about image: the artist must keep a contemporary look and sound in order to remain a good fit with current music. Pop culture moves quickly between trends, and it is important for the artist to maintain an image that doesn't change too quickly or dramatically, yet is still in sync with the times.

THE ARTIST'S SUPPORT TEAM

A basic set of professionals is needed to help the manager keep the artist's career on track, and it is the responsibility of the manager to assemble this group to support the artist. In recent years, the number of team members with nontraditional responsibilities has grown, as large labels have reduced the amount of support they provide to an artist's career.

Booking agent

The *booking agent* is the individual who connects an artist with most paid performances, and among the largest booking agencies in the world are the William Morris Endeavor Entertainment and Creative Artists Agency. A listing of some the other most active agents a manager will encounter on behalf of an artist would include the following:

AEG Live
Artist Group International
Bonus Management
Howard Rose International
International Creative Management
Live Nation Global Touring
Marshall Arts
MPL Communications
Paradigm
Rock Steady Management Agency
The Agency Group
Ujaama Talent Agency

Although agents book live performances, they are also involved in negotiating for artists to appear in commercials, arranging tour sponsorships, and for appearances in television specials. Agents generally do not get paid from the sale of recordings or from songwriting, although there are a few exceptions to this. In all cases, however, an agent is the person responsible for negotiating the fee an artist will charge a promoter for a concert performance or for an entire tour. Some larger booking agencies have what is termed a "responsible agent," who is the primary agent who receives offer letters from promoters and then creates proposed contracts for bookings for the artist manager's consideration. Artists are exclusive to agents for their performance bookings, meaning only one agent represents the artist for live performances.

The booking agent for major artists is required to coordinate available live performance dates with the manager and the record company in order to fill as many paying dates as possible—and to confirm that the artist is willing to commit to them. For their work as an agent, they are paid 10% of the value of the shows they book, so a $10,000 booking earns an agent $1,000. Some agents will reduce their fees for major acts and larger tours because even small percentages on booking a series of dates can generate considerable commissions for them.

As a practical matter, agents who book small venues may charge higher than 10% in order to make it worth their while to handle lower paying engagements. Typically, these small venues do not require the use of performance contracts required by the unions, although it is advisable to use union language to protect the artist.

Appearances on television talk shows are paid performances that do not involve the agent; they are instead usually handled by the artist's publicist and manager. This kind of performance is considered promotional in nature, and payments to artists who appear are minimal compared to fees charged to promoters for concert appearances.

Attorney

Nearly every activity the artist will have in the commercial side of the music business will involve promises in exchange for payment. How those promises are framed and how the compensation is planned become the heart of a contract, and the advice of an experienced attorney can assure the interests of the artist are represented in business agreements.

The attorney who becomes part of the artist's support team should be an entertainment lawyer and someone who is very familiar with today's music business. The attorney must be a specialist with a daily working knowledge of contract provisions, an understanding of the personalities with whom they will negotiate, a solid reputation within the industry, and experience deep enough to be familiar with most of the circumstances an artist will encounter during their career. Some artists have relied on lawyers who are not within the mainstream of the music business, and that has been adequate for them, but the best advice to protect an artist's career is to use the skills of one who is experienced and works it day to day.

Publicist

Publicists on the artist's team work for the manager; they are the conduit to news and information from the artist to traditional, satellite, and online radio, television, cable, newspapers and trade magazines—both online and offline, consumer magazines, blogs, and to all pertinent destinations on the Web. They have relationships with key information gatekeepers at relevant media outlets, and are trained and experienced in knowing the needs of editors and decision makers. They maintain databases of their contacts and understand how to get stories about artists placed in the media.

The major labels and many independent labels sometimes provide a publicist for the artist because they are interested in promoting a current recorded music project. But artists who are assigned a label publicist are sharing their attention with every other artist who has or is planning the release of a recording. As labels continue to reduce staffing to contain costs in a shrinking retail market, many managers now employ an additional publicist whose focus is only on the artist and who is under the direction of the manager. For a new artist who is in development, the manager will not likely choose to have a full-time publicist, but well-connected publicists are available who are willing to work on a project-by-project basis, thereby giving the manager a way to contain expenses in a budget yet have a publicist on the team.

Publicists charge a few hundred dollars to write and distribute a news release. For part-time, ongoing services for an active artist, they charge between US $3,000–$4,000 for three months, plus expenses. A full-time publicist becomes necessary when an artist releases a nationally distributed recorded music project, and costs for these services can be $4,000 per month or more. Another publicity-related cost is a photo shoot, and depending on the needs of the artist these can range from $500 to $10,000.

Manager of new media

New media actually aren't new, but compared to traditional media, they take on their own definition most often in terms of how people use digital formats to communicate and acquire information and entertainment. And because artist management firms have employed traditional media in their business relationships, "new media" as a team support function (as we define it here) has taken its place as a necessity to manage careers of artists. When the veteran owner of a management company was asked whether he planned to use new media within his company, he replied, "I don't understand it but I have no choice." This was his acknowledgment that he needed to use strategies using new media on behalf of his artists, even though it is an area in which he doesn't have expertise.

Enter the manager of new media. The most competitive artist management companies have a staff member who actively oversees the interaction of artists with their fans through their websites and through the various forms of social media. They oversee company content distribution for their artists, they track how users find and then interact with the company's and artists' websites, and they are the online eyes and ears for the artist brands managed by the company. A manager of new media requires a broad understanding of the music business and keeps current on trends in the use of digital media by the target market of the company's artists. For example, a new media manager should have quickly suggested the development of an inexpensive smartphone application, or "app," to keep fans regularly connected. To borrow from M. P. Godfrey, you can think of a manager of new media for an artist management firm as very much like a Swiss Army knife— one who understands fans and new media, and the tools to connect the artist and the fan with them (Godfrey, 2009).

Business advisors

There are four business advisors who could eventually become part of the artist's team. One of those is a banker. An artist will begin acquiring financial assets that require safekeeping, so a checking and savings account and a safe deposit box will be necessary tools. Likewise, artists will need credit cards for expenses on the road, access to lines of credit, and perhaps loans to help finance their career. Having a good relationship with a banker can assure that the artist will have access to these important financial tools. There are no direct charges by bankers for their advisory services, but they recapture their costs in the form of monthly account charges for credit, savings, and checking services as well as interest on loans and credit card balances. The global banking industry underwent sweeping regulatory changes resulting from the 2009–2010 recession, and it is important that the artist manager be aware how new regulations will affect the artist's business. Continuing updates are available at the website of the Federal Deposit Insurance Corporation (*http://www.fdic.gov*).

Another advisor an artist needs is someone who can recommend appropriate insurance coverage. Artists need liability coverage for their mode of transportation going to and from performances; they will need general liability insurance to be sure that neither they nor anyone associated with the live performance injures someone accidentally; they will need life insurance for their family; and they will need insurance for equipment that is taken on the road and is used for performances. Charges for these policies will vary depending on how much coverage is needed by the artist, and the cost of an insurance advisor is the commission he or she earns when he or she sells the policies. An advantage of membership in unions like the American Federation of Television and Radio Artists (AFTRA) or the American Federation of Musicians (AFM) is that they provide insurance coverage at a reasonable cost.

In the early stages of an artist's career, the manager may handle the routine collection of earnings, making bank deposits and writing checks to pay expenses. But as the artist becomes more active and earnings increase, it will become necessary to hire an accountant to handle these matters. Accountants who specialize in the music industry have experience that will keep the artist from developing liabilities such as income tax in other states or other countries, and they can take away the accounting task, thereby clearing more time for the manager to focus on career development. Charges by accountants vary depending on the needs of the artist. At the beginning of an artist's career, an experienced bookkeeper may be adequate to handle simple transactions and routine tax accounting, but as the career begins to mature, an accountant or a certified public accountant may be required, although their service charges are higher than the other options.

A business manager is someone who handles income and expenses of an artist and who ensures that what remains after the bills are paid is deposited or invested in ways that build wealth for the artist. Some business managers are accountants, some are certified business managers, and some are certified public accountants.

Whatever levels of business management the artist requires, it is important that the reputation of honesty of the business manager is solid, and that both the artist and the manager have unwavering confidence in that person's expertise. Artist management companies tend to have very small staffs, and often someone who receives money from the artist's commercial activity is also the same person who deposits the money, and that person may also be the business manager. In the author's experience, it is always advisable to have the office manager and the business manager be different people. Likewise, the person who handles the actual proceeds for accounts receivable should not be the same person who handles accounts payable, especially at small companies that have continuous income and expenses that vary, creating opportunities to hide theft by embezzlement. A small music industry company in Nashville had a full-time staff of four and the office manager was able to embezzle over $90,000 over a five-year period before her scheme was discovered. She went to jail for a period of time and continues to repay the money she took.

Just because your management company has a small staff, don't be tempted to save money by not separating the duties of those who handle and then account for the artist's earnings. An annual review of the company accounts by an auditor can add costs to the company but can help ensure the accountability of those whose duties include handling money.

Although there are other ways to compensate a business manager, those working in the music business typically charge up to 5% of the amount of money handled on behalf of their clients with a cap on the amount they may earn.

Other team members of artist management firms can include individuals responsible for radio promotion, marketing, sponsorships, licensing, and branding. These functions are mostly the result of the decline of support services by labels that the management company has had to assume on behalf of its artists.

ALTERNATIVE FORMS OF BUSINESS FOR THE ARTIST

Artists have options that determine the formal business type they assume. As these alternative forms are presented for basic informational purposes, it is important to remember that those who will help the artist and manager decide which business form to use will be the artist's attorney and accountant. It takes professional advisors to recommend a form that takes into account how the artist's business will be organized and also offers the most favorable tax consequences among the alternatives. The following is very general description of the alternative business forms available in the United States, and although other countries use similar terms as those described on the following pages, the application of local law to the terms may be very different.

Proprietorship

A *proprietorship* is the initial business form that a solo artist typically assumes. In this business form, individuals declare that they are in business for themselves, and they are their own bosses. They are not liable for anyone else's actions except those who they hire. It is an easy business form to create, and there is little regulatory accountability to the government. The key drawback is that there is unlimited personal liability when someone chooses to sue for accidents or negligence.

Partnership

This business form is the most common chosen by new artists who are actually members of a named group of performers. This occurs when a group of people pool their talents and professional resources to become "an artist," and a partnership is generally easy to form. Courts will support the right of any member of a partnership to act on behalf of all partners but will also acknowledge that all partners are liable professionally and personally for any lawsuits brought against them. Partnerships in Canada permit limited liability members who take no active role in the partnership but whose liability exposure is limited to what they have invested in the partnership.

Most groups beginning their careers do not have a partnership agreement, although it is very important to have one. Something as simple as a handwritten agreement by the partnership members will serve to forego some of the inevitable disputes that will occur between members. As an aid to groups who find themselves without a partnership agreement, Appendix C of this book includes an agreement form that gives guidance to bands or groups on forming a formal partnership. Contract forms suggested by this book or its website are for educational purposes, and are not intended to represent final contracts. In other words, always seek the advice of an attorney for the final details of the agreement.

Corporation

A *corporation* is a very formal business form. It requires filing a charter with a state's secretary of state, often with the assistance of an attorney, and it is regulated by government. It is easy to transfer ownership of the shares one owns, and it protects and shields the artist from liability because a corporation is its own legal business entity. It doesn't mean the artist will not be sued personally because of acts of the corporation, but it provides considerably more legal protection to the artist compared to a proprietorship or a partnership.

Limited Liability Company or Partnership (LLC)

Many times, a *limited liability company (LLC)* is a very attractive alternative for an artist or a performing group when compared to a corporation. An LLC operates and functions very much like a partnership but gives the members of an LLC the protection from personal liability in ways that a corporation does. Like a corporation, an application for an LLC is filed with the state's secretary of state and a charter is issued. Depending on the state in which the LLC is chartered, it can even have a single member (North Carolina Secretary of State, 2010). A variation of the LLC is the *limited liability partnership*, which carries many of the same attributes of an LLC. The limited liability partnership requires at least two members, and chartered LLPs in most states protect the individual members of a partnership from liability claims, although they are required to demonstrate adequate insurance and assets to settle any claims that might be brought against them. The United Kingdom has similar provisions, with their enactment of a law permitting the establishment of an LLP (HMSO, 2007); however, Canada does not recognize an LLC as a domestic business form, and use of an LLP in Canada is reserved primarily for professionals such as accountants and lawyers (Arvic, 2010).

THE INTERNET: A PRIMER FOR THE ARTIST MANAGER

The Internet is viewed variously in the music business as being either the technology that could ruin the recording industry or the entity that provides new opportunities for music creators and performers. However you view it, the relevant components of the Internet must be a part of the artist manager's promotional strategy for their clients. This section is included in this book to give artist managers a very basic understanding of key Internet components and the terminology that will be useful when directing the work of new media managers on behalf of their artist clients.

The importance of a domain name

A domain name is simply the registered name of the artist with a .com (or similar) suffix, and registering it should be one of the first things a manager does on behalf of an artist. The domain name becomes the address to the most valuable piece of real estate the artist can own on the Web. It isn't necessary to have a physical website created or space on a web server in order to reserve and register a domain name. For less than US$15 per year, an artist's name can be registered via sites such as Yahoo! and GoDaddy. Both services offer a free search service to see whether the artist's name is available. When the artist's name is registered, he or she then has a domain name reserved until the manager has the time and resources to arrange for the creation of the website for the artist.

Some artist names are easily misspelled by fans for visitors to a website, so it is always advisable to register all of the likely misspellings of the artist's name, too. When the webmaster creates the artist's website, they can cause the domain names with the misspelled derivations to be directed to the correct website.

What happens when someone registers a domain name that is the artist's trademarked name, and wants to charge a large sum to "sell" it to them? Can the artist be held hostage? There is a possible solution. The Internet Corporation for Assigned Names and Numbers (ICANN) can transfer the name to the artist, but it will be after there has "been agreement, court action, or arbitration" to provide them with terms of a settlement (ICANN, 2001). The resolution questions that ICANN considers are as follows:

1. Whether the artist's name is identical to that which has been registered by someone else
2. Whether the artist has a trademark that creates rights in the name
3. Whether the owner of the domain name has a reason to register it for his or her own legitimate business reasons
4. Whether the domain name was registered in bad faith, meaning that it was registered with the intent of making money through resale

In 2007, New Jersey graphic artist and designer Keith Urban (the artist) was sued by Capitol recording artist Keith Urban (the singer) for the right to use the domain name www.keithurban.com. The singer claimed he had the right to use the domain name because it is his trademark. The artist pointed out to the court that he had registered his domain name in 1999 and has used it as the address to his website where he offers his graphic design and art services. The artist does not want to give up the domain name, and the court as of this writing has not made a decision on whether the singer has a legitimate claim to it based on the ICANN guidelines. Meanwhile, the singer is using .net as the suffix to his trade name on the Web. When a decision is made, it will be posted to the website for this book at www. artistmanagementonline.com. Clearly, as in any issue that is as involved as this one, the manager should always seek the assistance of an entertainment attorney.

When registering the domain name for an artist, always try first to register their name with the .com suffix. There are other suffixes, such as .net and .org, but web users are likely to try the .com domain name first to find the artist. If the artist's name is not available, consider easy-to-remember alternative variations of a domain name that include the artist's name.

A URL

A *uniform resource locator (URL)* is the web address for the artist's website after it has been created and loaded onto a server. The artist has a domain name of "your-artist.com," which becomes a uniform resource locator on the Web when it takes on the URL or address of *http://www.yourartist.com.*

Web hosting services

Web hosting services are companies that set aside file space for the artist's website on their Internet servers. These servers are the physical location for the website, and it is from here that people using the Web get access to it. The cost of space on a server like this depends on the amount of file space required for the artist's website. For example, if the artist plans to feature large audio and video files, doing so will require a greater amount of space on the server. Costs for server space on hosting services is relatively inexpensive, so even larger files that showcase the artist provide more promotional benefit than their costs. Remember, too, that there is no geographic limitation on where the hosting service is located. Find one on the Web that has a good reputation and a competitive price—it can be located anywhere.

Some web hosting services say they are "free," but the price the artist pays is having banners and other advertising sharing each page. Using so-called free hosting services sends the message to visitors that the site and the artist are amateurs.

Content

Content for a website consists of all of the resources that are presented to the visitor to the website. Tom Hutchison published a list of the basic elements of an artist's website, which serves as an excellent guide for the artist manager to ensure that these key content elements are included on the site:

- A description and biography of the artist.
- Photos: promotional photos, concert photos, and other pictures of interest. This can include shots of the artist that capture everyday life, photos of the fans at concerts, and other photos that reflect the artist's hobbies or interests.
- News of the artist: press releases, news of upcoming tour dates, album releases, and milestones such as awards.
- Links that connect to the artist's presence on Facebook, Twitter, the label's homepage, Vevo, YouTube, and other important websites used by the target market.
- Discography and liner notes from albums.
- Song lyrics and perhaps chord charts.
- Audio files: these may be located on the purchase page.
- Membership or fan club sign-up page. Allows visitors to sign up for your newsletter, or for access to more exclusive areas of the site.
- Tour information: tour dates, set lists, driving directions, touring equipment list.
- E-store: page for selling recordings, T-shirts, and other merchandise.
- Contest or giveaways.
- Links to other favorite sites, including links to purchase products or concert tickets, venue information, the artist's personal favorites, e-zines, and other music sites. Ensure that all of your off-site links open in a new window, so the visitor can easily return to your site.
- Contact information for booking agencies, club managers, and the press.
- Message board for chat rooms: This allows the fans to communicate with one another to create a sense of community. This can be an area restricted to members.

- Blogs. A blog is simply a journal, usually in chronological order, of an event or person's experiences. Maintaining a blog of the touring experience is one way to keep fans coming back to the website to read the most recent updates to the journal. It also gives fans a sense of intimacy with the artist. Twitter is often referred to as a microblog because of the limited amount of text and links permitted in messages. Some artists discontinue Twitter because it can be a task trying to meet fans' expectations for continuous interaction. It needs to be a good "fit" with the artist's style in order to be an effective tool.[1]

This list does not include a specific reference to MySpace, but it is important to note that although MySpace has lost its place of prominence for general social networking, it continues to have appeal to those who offer and those who seek music. We noted earlier in this book of the millions of music acts that have a MySpace presence, and the site experiences traffic because the site has had a music-friendly format that draws both artists and fans of music.

A key feature that must be designed into the artist's website is a way to capture information about visitors. This is done by using a "form," which is nothing more than a page that opens on the site with spaces for visitors/fans to enter information that will be used to update them on the artist's activities such as upcoming CD releases or tour dates that have been added to their schedule. Capturing information about visitors is perhaps the most important function of an artist's website because it gives the manager and the record company the data necessary to specifically target people who have an interest in the artist.

Content on a website is simply a set of files that is organized by web design software. Creators of websites use *hypertext markup language (HTML)* to write code that generates web pages. There are also a number of software packages to create websites that are essentially drag-and-drop, meaning that they are developed and designed with mouse clicks and avoid the need to learn HTML. Among the more popular are Expression Web and Dream Weaver. The author maintains a website with links to sites and services noted in this chapter, a large free directory of artist management companies with hot links, and other information useful to an artist manager at *http://www.artistmanagementonline.com.*

Mining the Internet for information

Among the tactics used by labels of all sizes and many artist managers is visiting social networking websites to search for individuals who fit into the target fan base of the artist. Although it can be tedious, artist managers can find geographic information about individuals so they can email them when the artist will be in the area for a performance. They can also determine individuals' musical tastes that are similar to the music of the artist by searching these sites for similar artists, and

---

[1]There must be a commitment by the artist manager to ensure that the manager of new media gives the site regular updates and maintenance.

capturing contact information of people who are fans. Communicating about tour updates, new music, and other news becomes a part of the routine that draws people to the artist's website.

MySpace and Facebook were early social networking sites that opened the door for hundreds of other social networking sites with hundreds of millions of regular users. Consider all of these sites as potential sources for information about the specific target market of the artist.

An alternative to management seeking information to help build a fan base is having the existing people in an artist's database contact others regarding important information about an artist. Rather than the management company providing news in the name of the artist, the fans themselves can send news and recruit others to join the artist's database, sometimes referred to as "organically" building interest in an artist.

Artist managers who use strategies like those noted in this section give an advantage to artists in their client list, and can make them players in the latest incarnation of grassroots marketing.

References

Arvic. (1989). http://www.arvic.com/library/formofbusiness.asp.

Black, G. (2010). http://www.law.ed.ac.uk/staff/gillianblack_91.aspx.

CNN Headline News. (2007, January 15). *American consumers are exposed to 9,000 commercial messages each day.*

Franks, P. (2003). Designing Effective Websites. *Education To Go* Retrieved from http://www.ed2go.com.

Godfrey, M. P. (2009). *Social Media Community Manager Job Description.* Retrieved from http://www.aimclearblog.com/2009/04/05/social-media-community-manager-job-description/.

Hutchison, T., Macy, A., & Allen, P. (2009). *Record Label Marketing.* London: Focal Press.

HMSO (Her Majesty's Stationery Office). (2007). *Office of Public Sector Information.* http://www.opsi.gov.uk.

ICANN. (2001). *Uniform Domain-Name Dispute-Resolution Policy.* Retrieved from http://www.icann.org/udrp/udrp.htm.

North Carolina Secretary of State. (2010). http://www.secretary.state.nc.us.

US News & World Report. (1989, November 13). A U.S. Patent on Famous Voices. In *Thompson Gale* (p. 19). Middle Tennessee State University.

Income from live performance

Understanding and managing the sources of income for the artist are among the most important functions of the artist manager. Some artists want to be active participants in overseeing these aspects of their careers, but developing artists especially soon find out that the business side of their career is better handled by their manager. Until new artists become comfortable with someone else managing their earnings, it is be important that they participate in the financial accounting for their work, and, more importantly, develop the understanding of reports and documents that show their earnings and the costs it took to create those earnings. Depending on when they are ready to place major reliance on the manager to oversee their business affairs, they will meet regularly to review reports and financial summaries that will be as general or detailed as the artist requires. The sooner the artist finds the level of comfort in having the manager oversee their business affairs, the sooner the artist can build a focus on creativity and artistry, which are the elements that create the true value and earning potential in an artist's career.

Because disagreements over financial matters are often the catalyst for ending relationships, it is important that the artist be as much a part of the financial accountability reporting as is necessary.

The primary income sources for artists are various royalties from songs they have written, the sale of their recordings and related products, and the sale of tickets and payment of admissions at performances. Other sources of income are the sale of merchandise and the acquisition of sponsorships and product endorsements. Live performance is the income source that has the greatest number of elements that require the personal involvement of the artist manager, but live performance in the United States is an area of the music business that has thrived and grown even during a decade that included an attack on a major U.S. city, war, and the largest recession ever. The data from *Pollstar* in Table 8.1 clearly demonstrate that live performance is an area of the artist's career that holds opportunity though research shows a continued decline in the interest of young people in attending live music events (Webster, 2010).

Earning money from live performances is often the most immediate way that an artist can begin to support a career, and it is for this reason that it is so important for the manager to understand it and to exploit it for the benefit of everyone who is part of the business support group. It is an important way for the artist to build a fan base, to develop professionally, and to sell recordings.

There are five principal job areas that support an artist who will be earning from live performances: booking engagements, managing the tour, promoting the tour, managing the administrative business functions, and generally managing the artist. Figure 8.1 shows the relationship of these functions in the live performance hierarchy.

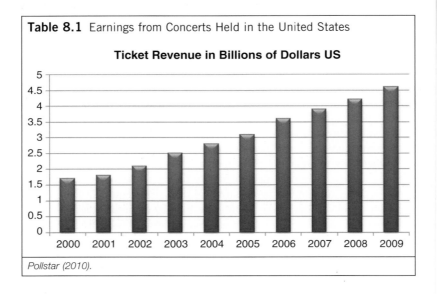

Table 8.1 Earnings from Concerts Held in the United States

Ticket Revenue in Billions of Dollars US

Pollstar (2010).

FIGURE 8.1

Concert Hierarchy

A manager of a small artist management firm often performs some or all these duties. In larger management companies, they are divided into functional areas of an artist's support team with individuals having these specific responsibilities.

There are two types of tickets that an artist will sell: soft tickets and hard tickets. Soft tickets are those that are purchased by fans to attend events where an artist is a participant with other artists at events like festivals and fairs and megamusic conferences like the Americana Music Conference or South by Southwest. From a management perspective, soft tickets help generate an occasional revenue stream, but for the long term, selling hard tickets is the objective. Hard tickets are those that fans buy to specifically see the artist. It's important that the artist develop a following that is interested in seeing them on a regular basis and willing to pay to specifically see the artist perform. The manager should make this a priority, because it creates an income stream that is specific to the artist and pushes the artist to build the confidence to become a headliner.

BOOKING THE PERFORMANCE

As you learned in Chapter 6, the agent is responsible for finding opportunities for paid live performances for the artist. There are agencies that specialize in booking performances in small clubs with 100–200 seats, as well as weddings and corporate events. Often they book scores of artists and are especially good for new artists or for those who do not want to rely on music to support their lifestyle. For the manager of a new client who needs seasoning in a live performance setting, small agencies like these can provide access to a range of smaller venues and performance settings that allow the new artist to take some chances in environments that frequently overlook mistakes. Commissions to agencies that book small venues and performances can be 20% or higher because the value of the performance contracts is relatively small. As is pointed out elsewhere in this book, agents franchised through the musician's union must handle bookings for larger performance opportunities for an artist, and their charges to book engagements for an artist can be as much as 10% of the value of the performance contract, although it sometimes is possible to negotiate a lower fee, such as when there is a series of performance dates with the same promoter.

When an offer from a promoter is accepted by the manager and the contract is signed, the agent collects a deposit or "guarantee" from the promoter with specific instructions regarding when additional payments must be made to the artist. For example, some performance contracts require a deposit with the signed contract and an additional percentage payment of the agreed amount 30 days before the engagement.

BUSINESS MANAGEMENT OF LIVE PERFORMANCES

Among the decisions artist managers must make is determining whether it is better to assign the business management of live performances to a business manager or to do it themselves. Either way it is handled, the artist manager must make a determination about whether an available paid performance or a series of tour performances will earn enough money for the artist to make the engagements worth accepting. The manager notes how much the performance will pay and subtracts all expenses associated with the engagement. The result, or the net, is the profit earned by the artist.

There are some performances that may have promotional value to the artist, and the manager and artist might agree that breaking even is a good result. Breaking even means that the payment received for the performance and the expenses associated with it are equal, and the artist makes nothing. An example of when a break-even performance might be practical is when an artist has the opportunity to open for a headliner, knowing that the appearance will be helpful promoting a recording, attracting the attention of radio, or building a fan base.

Budgeting for performances requires considerable planning by the business manager or the artist manager. A detailed listing of all costs of a live performance must be compiled. Overlooking one component of it could be the difference between making money or losing money, and the novice artist manager who does not have a strong background in bookkeeping or accounting should seek advice from an entertainment business manager. Anticipating expenses from live performances and budgeting for them is a skill that comes from the experience of doing it and living with the results. For performances that are booked with a promoter, it is typically the responsibility of the promoter to present a proposed budget to the artist manager through the booking agent before a contract is signed.

When creating a budget for performances that require tickets, it is important to consider what charges the ticket price includes in order to determine how much of the price actually goes toward paying the cost of the performance. Table 8.2 shows how a $25 ticket might actually be worth only $14 to the artist or the promoter *before* the bills are paid.

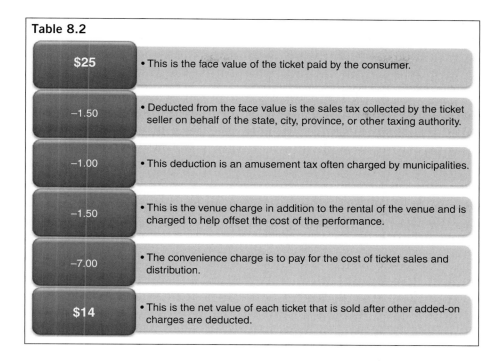

Table 8.2

$25	• This is the face value of the ticket paid by the consumer.
−1.50	• Deducted from the face value is the sales tax collected by the ticket seller on behalf of the state, city, province, or other taxing authority.
−1.00	• This deduction is an amusement tax often charged by municipalities.
−1.50	• This is the venue charge in addition to the rental of the venue and is charged to help offset the cost of the performance.
−7.00	• The convenience charge is to pay for the cost of ticket sales and distribution.
$14	• This is the net value of each ticket that is sold after other added-on charges are deducted.

So when estimating the income from the sale of tickets, it is important to use the net amount from total ticket sales, not the gross. Additionally, the budget must take into account exchange rate differences between countries, and to remember that "dollars" in the United States and Canada are not the same.

When an artist plays a ticketed venue that has different tiers of seating (the upper-tier "nosebleed" section versus the front row), it is important to have the business manager check the budget figures of the promoter to be sure that the projected income based on demand for each different ticket price is reasonable and accurate.

If the artist has a recording contract with a large record company, it is possible to negotiate an amount of money in what is called *tour support*. This is money that is given to the artist to offset losses artists incur while they are touring to promote an album for the company. When the budget is prepared, the manager presents it to the record company, noting how much money the artist will lose while on tour. The money the artist will lose is the basis for the amount of tour support money the label will advance to the artist. Tour support money may be offered by the record company in a range of $25,000 to $100,000, but in every case the artist must repay the money to the record company through recoupment. Although advances to an artist are typically commissionable for the artist manager, this is an advance that is not one the manager can draw from. Record companies view tour support as a promotional expense rather than a payment for services, and do not permit managers to draw commissions from them (Passman, 2009).

The current view of tour support by labels is that it is an investment with intended results in either the sale of recordings or getting the attention of radio. Tour support is no longer a standard commitment to an artist; rather, it is offered to artists when the label feels there is a potential for a "big payoff" and they sometimes will provide substantial sums to support touring (Dungan, 2007). New artists who open for established acts effectively earn nothing directly from their performances, so tour support becomes important to them. Likewise, tours planned for foreign countries often require tour support because of the added expense of logistics.

A budget for performances and tours should also include any amounts committed to an artist by a sponsor. Where it is possible, the artist manager should urge sponsors to provide their financial support before the tour begins. The early receipt of this income is helpful in building the show and assembling the physical assets that are necessary to accompany the artist on the road.

TOUR MANAGEMENT

The artist manager or someone on the manager's staff must handle the activities of managing a tour. As an artist becomes more active touring and the manager cannot be with the artist at every performance, it becomes necessary to hire a tour manager. This person is effectively an extension of the manager while the artist is on the road, a position that requires someone with patience, an orientation to details, and an understanding of human nature. The tour manager, sometimes referred to as a road manager, is the primary contact for everything while the artist is touring, and is totally responsible for getting the artist to a performance and ensuring that the performance is presented without any problems.

The tour manager takes primary direction from the artist manager, and then creates a detailed itinerary, beginning with the departure from the home base for the first show on the first date. When the artist manager has approved the itinerary, the tour manager begins finalizing the travel schedule, chooses and confirms the mode of travel, and reserves overnight accommodations. To organize each stop, the tour manager will use a template similar to the following list:

Date of performance	
City, state, province	
Name of venue	
Directions to the venue	
Time to load-in equipment	
Time for sound check	

Time doors for audience open to enter the venue	
Time the performance begins	
Length of the performance	
List of guests	
List of merchandise for the event	
Copy of technical rider	
Copy of the performance budget	
Catering for artist and other performers	
Confirm equipment rental	
Confirm hiring of stagehands and others to support the performance	

It's easy to see how an artist manager can quickly become overwhelmed by the necessary support for even a small tour. Each of the activities listed in the chart requires a number of telephone calls and emails, and follow-up is necessary to reconfirm them. This so-called advancing a performance becomes the guide for the person responsible for managing its success.

Among the other things the artist manager or tour manager does is meet with the promoter, or the person who hired the artist for the performance, during the show or immediately afterward to receive payment on behalf of the artist. This is called the *settlement meeting*. If the performance did not require a promoter, the payment terms are as they were specified in the contract, which is usually cash or certified check. If a promoter hired the artist, the manager or tour manager must compare the approved budget for the performance with the actual expenses and count the tickets sold for the event. The manager and promoter together total up the amount of money from the sale of tickets. Deducted from that total is either the approved promotional budget or the actual promotional budget, whichever is the lesser amount. The result is the profit from the performance.

PROMOTING THE PERFORMANCE

Unless a performance is in a small venue and is being used for developing stage presence or seasoning the artist as a performer, an artist's performance or tour appearances require very active promotion. This is achieved through publicity, advertising, and an active presence on the Internet, especially through social networks. However, it depends on the career status of the artist whether all of these promotional avenues will be available.

A new artist without a contract with a record company will not be able to afford much advertising to support a tour. Artists in this category and small independent labels have very limited financial resources, so there is heavy reliance on publicity and social networking to promote an artist's performance and to have it covered by local media when the show dates arrive. The artist manager works closely with the artist's publicist to maximize potential local radio, television, and newspaper publicity. The webmaster or new media manager for the artist must keep the artist's website current and social networking sites buzzing with the most recent tour schedule, links to the venues, information about tickets or admission charges, and should be sending occasional emails to the artist's contact list about additions to the tour schedule.

An artist who has an established fan base and has had several recordings released in the past will find it easier to get publicity to support a performance or a tour. Simply releasing a recording on an independent label will give the media something to write about and generate a new interest in the artist. The new recording, coupled with a tour and a unique story angle, can create a series of stories for the media, as you will see in Chapter 10.

For the artist signed to a larger label, the promotional efforts that will directly affect the sale of tickets include radio airplay and video plays on cable television and on Internet sites, especially in markets where the artist plans a tour stop. Labels will also assign a publicist to advance the scheduled performances. To the label, this kind of promotion helps them sell recordings while immensely helping the sale of performance tickets.

Strategically, the artist manager should consider employing some of the label's promotional tools in places and ways the record company is not planning on spending money or effort. For example, some labels create videos of singles that have been released to radio, but they promote the use of the video only on major cable music channels and websites that feature music videos. In the United States, there are hundreds of independently produced video programs on local cable systems that are easily convinced to show music videos. For less than US$4,000, a manager can hire a video promotion firm to promote the artist's music video to these local video shows for 12 weeks or for the length of the time the single is on the chart. The artist manager who uses the already-produced video in this way will be able to promote the artist to video outlets that the label does not promote to, and it will serve as a catalyst to sell tickets in smaller markets as the artist tours. This makes especially good sense considering that the artist paid for half, if not all, of the cost of producing the video through recoupment. With artist recording contracts now including required sharing of performance income with the record label, it may become easier to enlist the label's participation in financing video promotion to smaller markets.

The promoter

As the artist becomes well known and builds a sizable fan base, the artist's booking agent will seek concert promoters who are willing to hire the artist for a performance in a particular market or for a series of performances on tour. The promoter is an entrepreneur. He or she pays all of the costs of producing and marketing a live performance, and as a result also takes all of the financial risk.

To minimize their risk, promoters purchase insurance on shows they produce. If an artist becomes ill and cannot make a committed performance date, the promoter will lose the money invested in promoting that date and must refund money to ticket holders. All of this can create an insurable loss to the promoter. An example is the exposure concert promoter AEG Live had with the untimely death of Michael Jackson just before his London concerts were to begin in 2009. The company had given Jackson a $10 million advance to agree to the performances, and paid another $30 million to produce the show for its theater performances. The insurance policy AEG purchased included a physical of the pop star, which showed him to be in good health, and it was intended to cover potential losses in the event the shows were cancelled. Without an insurance policy, AEG would have no way to recoup its losses from selling tickets and producing a show that never materialized. It was anticipated that the Jackson estate would return the advance but it is unknown whether the insurance policy was large enough to cover all of AEG's losses (Mayerowitz, Tadena, and Gomstyn, 2009).

For each performance, the promoter creates a budget, leases the venue, contracts with performers, hires local labor, leases all necessary production equipment and expertise, purchases all permits, pays all taxes and fees, pays for all marketing expenses, and supervises all aspects of the event. The specifics of all of these activities have the prior approval of the artist manager, and the tour manager has the final say on behalf of the manager on the day of the show.

When the show is over, the artist is paid the guarantee for the performance and the promoter is paid, for example, 15% of the net proceeds. This then becomes what is called the *split point*, which is the remainder of earnings from the performance. It is the amount at which the promoter and artist have agreed that they will begin sharing at (for example) a ratio of 30% to the promoter and 70% to the artist. Agreements with promoters also include a net earning amount termed *overage*, which is the amount at which the artist takes all of the remaining excess earnings. It is easy to see that established acts require fewer services from a promoter because of the public awareness of the artist as a brand, but the inclusion of some sharing of excess earnings will help ensure that the promoter will meet the attendance expectations of the artist and the manager. However, the inclusion of a generous split point and overage with a new artist adds important incentives to the promoter. And there are circumstances in which an artist prefers to accept only a guarantee for their services, which may occur when the artist is dealing with an unfamiliar promoter or with one who has a history of high expenses.

When an artist manager considers contracting with a promoter, some care must be taken and the manager should look to the agent for advice if he or she has never used the promoter before. Some of the things the manager and agent should consider include the following:

- Does the promoter have a good overall reputation?
- How long has the promoter been in business?

- What is the promoter's general success record, especially with this kind of an artist?
- Is the promoter strong in markets where the manager is planning performances for the artist?
- Can the promoter handle a complete tour?
- Does the promoter have good relationships with concert production service companies?
- Does the promoter have the financial resources necessary?
- Will a record label be part of the decision regarding the promoter?

Even for promoters with whom the manager has had some experience, occasionally revisiting some of these questions about them can prevent surprises that will have unpleasant consequences.

THE PERFORMANCE CONTRACT

Live performance agreements contain basic information such as the name of the company or person hiring the artist, the date and time of the engagement, the specific location of the performance, the time and length of the performance, the services the artist will provide, and any riders to the contract. A *rider* is an attachment to the main portion of the contract that specifies standard production requirements for the artist's performance as well as any personal needs the artist has prior to and during the performance. Riders include very technical specifications for production crews to follow when preparing the stage for the performance. Riders can also include things such as limousine service for the artist, elaborate food before the performance, expensive beverages, and luxurious dressing room accommodations. For major artists on tour, concessions of this type are often justified as a way to reduce the rigors of being on the road for lengthy periods of time. The new artist can expect small but reasonable personal concessions in a rider. However, all expenses associated with any rider are charged directly back to the artist as costs of the artist's performance, and are deducted as an expense from the receipts from ticket sales. This is an important consideration for the artist manager or business manager as they create a budget for a performance or tour.

The performance contract or agreement includes the circumstances under which an engagement may be cancelled. Those reasons include the death of key personnel such as a band member, the tour manager, or production staff, and the health of the artist. If there is the possibility that the artist will be exposed to personal danger—for example, because of threats—the performance may be cancelled. Under all of these circumstances, the artist manager must return any deposits that have been paid. For the promoter, individual, or company hiring the artist, however, there generally is no recovery of their losses resulting from the causes for cancellation provided for in the contract unless they have purchased insurance to cover the potential losses.

The artist manager should include provisions in the contract that give him or her the right to approve advertising for the engagement and proof that the advertising was actually bought. The manager will also want approval of the distribution of all complimentary and promotional tickets to ensure, for example, that all radio stations in the artist's musical format in the market are treated equally. As the manager reviews the list of people requesting premium complimentary seating, they should know that artists can become upset when people in the first two rows of a concert audience are people who are "suits," or people who were given priority seating but aren't the energized fans the artist would prefer to see in the front rows. A good rule of thumb is to make it a policy that priority seating for required complimentary tickets begins behind row 5.

If the artist is the headliner for the performance, the manager should include language in the agreement that gives the artist and the manager the approval for any other artists who may appear on stage as opening acts. Other sections of the contract are specific about details of the performance, how licenses and permits will be handled, and how settlement and payment will be handled.

MERCHANDISE

Selling artist's merchandise at performances and on the artist's website is another way live performances can quickly begin generating income to support the artist's career. If the decision is made to offer merchandise for sale at performances, the artist and the manager must make an early commitment to purchase only the highest quality items they can afford for sale. Cheap merchandise makes a statement about the artist, so it is better to offer nothing for sale rather than something that is of poor quality.

It is the manager who arranges the production, manufacture, and sales of merchandise. However, as a new manager will quickly learn, this income stream can become extremely time-consuming. Among the steps in handling merchandise for the artist are:

- Creating a budget for merchandise
- Designing each item of proposed merchandise inventory
- Determining quality
- Determining quantity
- Ordering finished art
- Determining pricing
- Estimating obsolescence
- Getting bids from manufacturers
- Arranging manufacture
- Arranging shipment from manufacturer
- Arranging storage
- Purchasing insurance

- Creating and maintaining an inventory
- Packing and shipping merchandise to the location of each performance
- Paying hall fees to the venue to be able to sell the merchandise
- Arranging space and people to sell the merchandise at performances
- Arranging management of the sale of merchandise at performances; accounting, inventory, cash management, reports
- Packing remaining merchandise after performance and shipping it back to storage

One budget item the manager should be prepared to see is something called *hall fees*, sometimes also called *house rate* (Barnet, Waddell, and Berry, 2007). Many arenas and other venues charge the artist as much as 25% of merchandise income from a concert appearance as a fee for the privilege to sell merchandise at a venue. As is often the case, major artists have more negotiating power to be able to reduce those hall fees for merchandise sales because the venue will still be able to sell a large number of $8 slices of pizza and $6 soft drinks.

Because of the time merchandise management can consume, managers often see the value of licensing the image and marks of the artist to a reputable merchandise company and making it their responsibility to create and sell merchandise at the artist's performances. Merchandise companies pay onsite costs as well as licensing royalties to the artist ranging between 30% and 40% of the gross receipts, meaning that a $25 T-shirt will generate over $8 to the artist for each one sold.

Bootleggers are rogue merchandise sellers who appear near venues the day or evening of a performance and offer counterfeit merchandise for bargain prices. Because an artist has the right to their image in commerce through trademarking or service marking, the manager is responsible for alerting local authorities that people are selling illegal merchandise and press to have them arrested. This has been a problem that is literally difficult to police; however, many managers, promoters, and venue operators have had increasing success in stopping counterfeit sales with the assistance of local authorities.

Negotiating licensing contracts with merchandise companies can become somewhat complex; this is another area where the artist manager should be especially careful. The merchandise company will be expected to pay a royalty advance to the artist, and will want guarantees in the licensing contract of the minimum audience sizes at upcoming performances. The merchandise company is interested only in the number of people who actually pay to be part of the audience, feeling that those who receive free tickets are not closely enough connected to the artist to want to buy merchandise. If the audience estimates are too high, the contract will require the repayment of money advanced to the artist, and the artist may lose all of the benefits of licensing to the company.

The manager may also want the merchandise company to provide fulfillment of orders made from the artist's website. Fans who click on merchandise on the artist's site will be redirected to an order form at the merchandise company. Here, the company collects the payment, packs, and ships the merchandise to the buyer. The merchandise company provides these services to the artist for a fee, but for the manager who is not prepared to offer fulfillment services, this can be a real time saver. If merchandise is handled in this manner, it is important that the artist's webmaster maintain high-quality and current images of the items being offered for sale. These points are key determinants of the appeal of the merchandise to visitors to the artist's site.

When preparing to offer a merchandise licensing agreement to a company, it is always advisable to seek the advice of an entertainment attorney who is experienced in this specialized kind of an agreement.

A final note about live performance is to make the artist manager aware of *Pollstar* magazine as a planning and information resource. *Pollstar* is the trade magazine of the concert and touring industry that supplies weekly information about artists, venues, promoters, audience sizes, ticket sales, and gross earnings of live performances for hundreds of acts. It can help the artist manager with budget estimates, the timing of a tour, and will help identify agents and promoters who are successful with artists similar to the manager's clients. The free online version of *Pollstar* gives information that is more useful to fans than to management. The actual magazine and the online subscription service offer the depth of information a manager will need.

INTERNATIONAL TOURING

Plans for international touring are often created to build a following in other countries before launching a career domestically, or planned at the end of a tour supporting an album that has nearly reached the end of its prime life cycle. One of the most famous groups that used international tours to launch a career in the United States is the Backstreet Boys. Manager Lou Pearlman put the group on international tours from 1994 to 1997, preparing them for the eventual acceptance by the fans of pop music in the United States (VH1, 2005). Successful international touring for established artists almost always requires a successful album release at home, and labels will be reluctant to support an extended international tour until the domestic market has been served.

Touring in other countries has become more challenging after the September 11, 2001, attacks in New York. Required paperwork can now include a passport, a visa, a work permit (in Canada), an international driver's license, copies of signed contracts, a bond to assure customs officials that the artist will not sell any of their equipment while in the country, certification that travelers are healthy and have the necessary immunizations, a detailed equipment manifest, insurance policies, and a carnet.

A *carnet* is essentially a passport for the equipment being transported to another country for use by artists in their performances, and its use can expedite getting the equipment through customs in 75 countries. It also prevents tours from being required to pay import duties or taxes on equipment used for performances that will be leaving the country with the artist (Barnet, 2006).

Managers of artists who travel internationally for touring or for pleasure must have a passport. It is the primary document required to permit individuals to enter into other countries, and to return to their home country. In January 2007, the United States began to require all international air travelers to have a passport when landing—even when traveling from border-friendly countries like Canada. However, a passport card, which takes less time to apply for and to receive, is intended for visitors to countries that border the United States but who are not flying. The manager should plan ahead by at least six weeks from the time a passport is needed for the artist and touring personnel to travel to its delivery, although the U.S. Department of State offers expedited passport processing that can be completed in two to three weeks (Department of State, 2010).

In this time of using online services to schedule travel, the services of a travel agent can be helpful when planning a tour. Some of the larger agencies that serve the music industry are aware of travel requirements to many destinations, and can be extremely helpful with occasional and onerous changes in itineraries. Other useful sources of information include the websites of embassies, offices of the American Federation of Musicians, AAA, and online services from the U.S. Department of State.

American tour promoters who have international stops usually associate with promoters from the country into which the tour will go. Likewise, an artist's agent in the United States will likely have agents who handle international bookings for them, and they often have offices in the United Kingdom that assist in coordinating international tour stops (Barnet, 2006).

A final word of caution to the manager when planning international performances: be aware of the income tax regulations in the countries visited. For example, a series of tour stops in Canada requires that the company or individual who pays for the artist's services withhold 15% of the gross earnings for each performance (Canada Revenue Agency, 2010). Likewise, international artists will be required to pay income tax for earnings in the United States, with the largest international tours having an agent of the U.S. Internal Revenue Service assigned to track the earnings of the tour.

COLLEGE TOURS

College touring can be valuable for a manager who is helping an artist with building a fan base composed of this target market. Often engagements barely pay for the costs of the appearance but can add valuable experience for newer artists who are trying to learn about themselves in a live performance environment, and they can help build fan bases.

Managers will quickly find that most campuses have committees that have a fixed budget for the school year and plan appearances by artists based on available funds. It also means that there is a group of people making the booking decision rather than a single promoter or agent. Like most business that is conducted by committee, it is not quite as prompt as when there is one decision maker, so the manager should provide additional planning time, knowing that contract offers will be slower to arrive.

Some colleges use what is called a *middle agent*, whose job is to negotiate with managers or agents on behalf of the college committee. These agents work for the college, and can be helpful to managers who do not want to deal directly with the college (Ostrowski, 2005). Other useful information for the artist manager considering a college tour is available through the National Association for Campus Activities. The organization conducts regional and national meetings and seminars that include opportunities for artists to showcase for college campus activity committees and middle agents.

References

Barnet, R. (2010). Personal interview.

Barnet, R., Waddell, R., & Berry, J. (2007). *This Business of Concert Promotion and Touring.* New York: Billboard Books/Crown Publishing.

Canada Revenue Agency. (2010). *Individuals: International and Nonresident Taxes.* Retrieved from http://www.cra-arc.gc.ca/tx/nnrsdnts/ndvdls/menu-eng.html.

Department of State. (2010). http://travel.state.gov/passport/processing/processing_1740.html.

Dungan, M. (2007). Personal interview.

Mayerowitz, S., Tadena, N., & Gomstyn, A. (2009, June 26). *Jackson's Death Means Multimillion-Dollar Woes for Concert Promoter.* ABC News. Retrieved from *http://abcnews.go.com/Business/MichaelJackson/story?id=7938179&page=1.*

Ostrowski, D. (2010). *Booking College Shows.* Retrieved from http://www.starpolish.com.

Passman, D. S. (2009). *All You Need to Know About the Music Business.* New York: Simon and Schuster.

Pollstar. (2010). January 18, p. 5 Year-end concert data.

U.S. Department of State. (2010). *International Travel.* Retrieved from http://travel.state.gov/travel.

Vasey, J. (1998). *Concert Tour Production Management.* Burlington, MA: Focal Press.

VH1. (2005). *Behind the Music: Backstreet Boys.*

Webster, T. (2010). *The American Youth Study 2010, Edison Research.* http://www.edison-research.com/home/archives/2010/09/the_american_youth_study_2010_part_one_radios_future.php.

Income from songwriting

Artists who are also songwriters, or who can become songwriters, add a valuable income stream to support their careers. When a songwriter writes a song that is recorded, the writer is entitled to earnings in the form of royalties when a song is performed live, on the Internet, on radio and television, in a dance club, and each time it is sold as one of the songs included on a recording. The songwriter also earns royalties from the printing and publishing of sheet music, karaoke tracks, ringtones, and from licensing the song for use in advertising, and in video game, movie, and television soundtracks. Clearly, there is considerable potential income for the artist who also develops into a good songwriter. And finally, songwriting can ensure that artists will have material they can perform that was written for themselves as part of their total creative expression without necessarily relying on other writers.

COPYRIGHT

For the artist manager with a limited background in songwriting, a discussion about income from songwriting begins with an understanding of copyright, which establishes the rights of a writer to protection from theft and is the basis for ensuring earnings from creative works.

Copyright is the right a songwriter has to his or her own work after it has been put into tangible form, typically meaning that it has been written down on paper or saved to some electronic medium like a hard drive or flash drive. Once it is in tangible form, it is exclusively the right of the songwriter to allow it to be copied or used for any purpose that they authorize. It is a common understanding that a song must be registered with the Copyright Office before it is copyrighted; however, registering the song merely adds more protection from predators who try to claim rights for themselves from the original writer's work. Once the song is written down, it becomes copyrighted. Registering a U.S. copyright online is as easy as going to *http://www.copyright.gov/forms/*, filling out a form, and paying a fee of $35. If you prefer to handle the registration by mail, the Copyright Office charges $50. Understand that it may take up to eight months for your copyright to become registered, regardless of which method you choose to register it.

For purposes of recording a song, a writer holds the right to grant permission for someone to be the first to record that song. The first recording of a song by a recording artist on an album is granted what is called a *mechanical license*, which tells a record company what it must pay the songwriter for each copy of the recorded song that is sold. Once the copyrighted song has been recorded, the songwriter is required by law to issue what is called a *compulsory license* to anyone who wants to record the song. Original song licenses usually require the payment of 9.1 cents for the sale of each copy of a recorded song; compulsory licenses for previously recorded songs that are less than five minutes long require the payment of a statutory rate (meaning prescribed by law) of the same amount as an original mechanical license. Longer recordings require higher license payments. The statutory royalty rate typically increases every few years.

Earlier versions of copyright law protected songwriters for varying lengths of time, but the 1998 amendment to the U.S. Copyright Act says that a copyrighted song that was written after 1978 enjoys the protection for a term of the life of the author plus 70 years. For many songwriters, this assures that they will earn from their creative work for their lifetime and for the lifetime of those who benefit from their estate.

Copyright law provides for the payment to songwriters whose works are used in digital formats. The Digital Millennium Copyright Act of 1998 created a payment mechanism for the use of a songwriter's works on the Internet and on satellite radio.

Managers will find that their artist/songwriters also enjoy the protection of their works by most foreign countries. The Berne Convention in 1989 resulted in an agreement to protect American copyrights in all of their member countries.

And finally, an artist manager should know the things that cannot be copyrighted: the titles of songs, songs that have been written but haven't been fixed (placed into a tangible form), names, ideas, and information (Rolston, 2008). Volumes have been written on the subject of copyright, but as an extremely basic primer, this chapter is a good starting point to discuss songwriting and publishing. Typically, a songwriter will sign a contract with a publishing company agreeing to transfer their copyrights to the publisher in exchange for their promise to exploit the copyrights (arrange for the songs to be used commercially) for the benefit of both, which is where we begin.

SONG PUBLISHING

Artists today must often be the "complete package" if they are to have a genuine opportunity at a lasting career in the music business. They need to be skilled at performing, playing an instrument, and writing songs. If the manager signs an artist who is not a songwriter, it is advisable to begin nurturing that talent to see if it develops into a potential income stream.

What are some of the qualities that increase someone's potential to become a hit songwriter? An informal survey by ASCAP asked current publishers of hit songs what attributes songwriters should have if they are to become successful. The top five were:

- Ability to write great lyrics
- Personality/compatibility with the company
- Ability to write great melodies
- Ability to write alone
- Affordability (amount of draw)

The "amount of draw" (Murphy, 2010) refers to how much the publishing company must periodically pay a writer until their creative works are recorded and begin earning royalties. Among the easiest ways to begin developing the artist into a songwriter is to get the artist a writing coach, and then have them begin cowriting songs with experienced writers. It may prove that the artist will not be able to become a viable songwriter, but the manager cannot overlook an opportunity for the artist to grow creatively as a writer.

Both the manager and artist should understand that if publishing income is commissionable by the artist manager, the draw received by a songwriter is also commissioned. A draw is considered advance payments against future royalties and is recoupable under most circumstances.

If artists do not have their own song publishing company, they need to become associated with a publisher. A publishing contract is very much like a recording contract, in that the publisher agrees to exploit the commercial value of the artist–songwriter's creativity. A publisher is someone who becomes a partner with the writer: The songwriter agrees to write songs exclusively for the publisher and the publisher agrees to handle the administrative part of publishing those songs and getting songs placed with record producers who might have an interest in songs for artists they are producing. For artists who write only for their own recordings, it is often convenient to have a publisher handle administrative things such as registering the copyright and collecting royalties for them.

The standard arrangement for a songwriter who signs with a publisher is to give the copyright to the publishing company and receive 50% of the publishing royalties. The other 50% is given to the publishing company in exchange for getting the songs placed on recordings and in other commercially worthwhile endeavors. If a songwriter has cowriters, the writers' 50% will be divided between them based on their agreed sharing of the royalties. For example, three cowriters might agree to share one-third each of the writers' half of the royalties earned through song publishing.

The flow of earnings for mechanical licensing of songs from the publisher to the artist–songwriter is quarterly, and is based on the cycle of payment from an organization called the Harry Fox Agency. This company licenses songs on behalf of many U.S. song publishers and then collects the royalties for their use to be paid to the publisher. The strength of the Fox Agency is in its ability to periodically audit record companies to be sure they have paid their publishing clients royalties that are owed, and it means that small publishers do not need to deal with licensing issues. The Canadian counterpart to the Fox Agency is the Canadian Mechanical Rights Reproduction Agency (CMRRA, 2010).

INCOME FROM SONGWRITING

The first time a song is scheduled to be recorded, the record company must pay a mechanical license to use the song. This is a totally negotiated license fee that is paid to the publisher of the artist's song in the form of royalties and split with the writer(s). There is no limitation on what the license charge may be. For example, it could be negotiated to cost the label 11.2 cents or 9.1 cents or any other amount for each copy of a recording that is sold. However, after the song has been recorded and released commercially and begins appearing on other albums, it will earn the publisher and songwriter 9.1 cents for each album that is sold (U.S. Copyright Office, 2010). Songs longer than five minutes have a higher statutory rate. This is the current statutory rate a publisher may charge for a compulsory mechanical license for the use of one of the songs in their catalogue (song inventory), although a lower rate may be negotiated. In late 2008, the US Copyright Royalty Board approved a 24-cent royalty rate for each ringtone sold.

Mathematically, if an artist's song appears on another artist's album that sells 20,000 units and the compulsory rate of 9.1 cents is applied, the publisher receives $1,820. Assuming there is a 50/50 split between the publisher and the songwriter, the artist and the publisher receive $910 each. If the artist has a cowriter or if a preexisting sample is part of the song, the writer's amount is reduced further. Keeping current with the budget for the artist is a continuing management responsibility, so to predict the amounts and timing of income from songwriting, the manager must rely on the best estimates of the publisher and the record company.

If the artist is the writer of songs that appear on his or her own album, called *controlled compositions*, most record companies will offer to pay only three-quarters of the full statutory rate and only for the first ten songs that are recorded. This means the label pays 6.825 cents per song on the album in which the artist is also the songwriter. From the record company perspective, they feel they are promoting the recordings of the artist–songwriter from which they will be earning royalties for the recording. Because the label is doing the promoting, the writer of the music is also getting a benefit, and that benefit should be shared with the label by reducing songwriter payments and thereby reducing their cost of sales.

If the number of songs exceeds ten, the label will charge the artist for royalties for the excess songs. Even if the artist has written the songs, remember that the royalty payments are made to the publisher, who is a partner sharing earnings with the artist–songwriter. Limiting the album to a specific number of songs limits the number of compositions for which the label must pay a copyright fee each time an album is sold, again. If the artist wants twelve cuts on the album, the label will require the artist to pay royalties for the additional two songs.

For artists who publish for their own publishing company, some of the income considerations just noted still apply. The primary difference is that there is not a split with the publisher. Even if the artist also owns the record label, the manager must remember that some songs on an album will be cowritten or written by other members of the artist's group, which will affect costs and earnings for members of the group.

Timing of publishing income, as noted previously, becomes an important consideration for the artist manager who must manage budgets for their artists. Melissa Wald, owner of Copyright Solutions, provides this guidance on publishing income from songwriting:

> *For both the new songwriter and the seasoned songwriter, keep in mind that most labels take anywhere from two to four quarters to begin reporting sales/royalty earnings to the publishers. Many times a mechanical license is not in place within the first six months of a release. That being said, most publishers report to songwriters on a semiannual basis, 90 days after the close of the period. So, let's say an album was released in September . . . even though that is in the third quarter, the label most likely is not going to actually report those sales until the fourth quarter or first quarter accountings. We could assume you are lucky and they are reported in the fourth quarter sales statements. Those statements are due to be issued and paid on February 15. Because the publisher is on a six-month reporting schedule, it will be September 30 before those sales/earnings are reported and/or paid to the songwriter. That is almost one full year under the most optimistic timeline. If there is a radio single, it will be nine months to a year for the songwriter to start seeing the bulk of that income as well.*

(Wald, 2010)

INCOME FROM SONG PERFORMANCE

For the artist–songwriters who have success with their music appearing on traditional, satellite, and Internet radio, and other public performances of their songs, earnings from the performance of their compositions can provide a nice addition to the income stream. Companies called performing rights organizations (PROs) license the use of copyrighted musical compositions for public performance and then collect royalties for the use on behalf of publishers and songwriters. A PRO does not license the use of songs in dramatic productions, for example in movies and on television, or in commercials; those licenses are negotiated directly with the publisher, for which there is no established fee. In the United States, BMI, ASCAP, and SESAC are the major PROs; SOCAN and Resound are among the Canadian PROs; and PPL is among the well-known performance rights organizations in the United Kingdom.

SoundExchange functions in the United States like a PRO, in that it collects performance royalties on behalf of the owners of copyrights of sound recordings used on satellite radio and on the Internet. The owners of copyrighted recordings are typically record companies, and SoundExchange secures payment on behalf of the label as well as for those who perform on recordings such as artists and their musicians. Traditional terrestrial radio (stations with broadcast towers) in the United States has been exempt from paying artists and labels for the use of their recordings, but there is an active movement to reverse the exemption.

If an artist–songwriter is not registered with one of the PROs, he or she will not earn income from the performance of songs he or she writes until he or she registers. The artist manager must be sure the artist who becomes involved in songwriting is registered with a performing rights organization. It costs nothing and can be done online in just a few minutes.

The amount of money paid to a songwriter through public performance is based on formulas that have a series of variables that include primarily the number of exposures of the performance of a song to an audience, its chart position on radio airplay charts, and how much money the performance of the song is making for the person or company that is using it in their business. The audience for the writer's song, for example, can be in an amphitheater, on radio or television, in a restaurant, in a retail store, on the Internet, on satellite radio, and in an elevator.

Payments for most of the licensed uses for the performance of songs are made by the PRO directly to the songwriter and to the publisher at the same time, on a quarterly basis. The manager should note that unlike the sale of a song, where payment comes through the publisher, performance income comes directly from the PRO to the writer (Hull *et al.*, 2011, 132).

Because of the number of variables that are used by PROs to determine the value of performances they license, it is difficult to say how much performance income is. Donna Hilley, former CEO of Sony/ATV, estimates that the performance income resulting from an active number-one country song would be an estimated US$500,000 (Hilley, 2008). All of the 10,000 commercial radio stations in the United States pay fees to PROs in order to use a songwriter's music on the air, and it can amount to a substantial sum. Certainly, every song will have its own value in performances, but it is clear that adding songwriting to the skill set of the artist holds considerable potential in career earnings.

When you consider the various sources of income for an artist/songwriter who wrote a song for someone else, a theoretical but realistic example of how much a writer could earn from a country song that reaches the top of a sales and airplay chart is shown in the following table:

Royalty Source	Amount
Mechanical	$38,675
Print	$500
Foreign	$2,500
Ringtone	$3,000
Karaoke	$300
Synchronization	$1,250
Performance	$500,000
Total	**$546,225**

Source: CPA Craig Owens, 2008; assumes 750,000 albums sold, 100,000 singles downloaded. Figures are calculated after split with publisher.

As noted earlier in this chapter, payments to a songwriter—unlike payment for a concert performance—can extend two to three years, which becomes a budgetary consideration for the artist manager when considering timing of income.

PUBLISHING AS A NEGOTIATING ASSET

If the goal of the artist is to record for a large label, the ability to write songs becomes a tool that the manager can use to negotiate a recording contract. Most major labels and larger independent labels have publishing companies associated with them that become home to their artists who also write music. With the continued erosion of the sale of recorded music, labels look for buffers from these losses by acquiring the rights to publish the music of new artists who are signed for recording deals. If the publishing of the artist is not inextricably tied into another publisher, the manager can bring the artist's publishing to the negotiating table with the record label when pursuing a contract with the company. There will also be instances in which the artist is unwilling to give publishing rights to the record label, so the manager will negotiate without it as a bargaining point.

Finally, if the artist is a songwriter and is seeking a recording contract, the manager must coax the artist to consider using the songs of other songwriters on his or her albums if better music can be found. There will be a temptation to use only the artist's songs on the album, but the competition to get music the best placement among the vast electronic/digital media means the artist must record the best songs there are. The best songs get the most exposure to listeners and viewers of music videos. Do not let the artist accept anything less than the best music available, despite what the producer and the label's A&R department might want. This is the perfect point to begin looking at another major source of income for the artist—recordings.

References

CMRRA. (2010). http://www.cmrra.ca.

Hilley interview, circa 2008.

Hull, G. Thomas Hutchison and Richard Strasser. (2011). *The Music Business and Recording Industry,* 3rd ed, Routledge, Ny, NY p132.

Murphy, R. (2010). *What Publishers Really Want.* Retrieved from http://www.ascap.com/nashville/murphy/index.html.

Owens, C. (2008, October/November). Writer Earnings from a No. 1 Country Song. *Music Row Magazine,* pp. 48–49.

Rolston, C. (2008). *Lecture notes from Belmont University.*

U.S. Copyright Office. (2010). http://www.copyright.gov.

Wald, M. (2010). Personal interview.

Income from recording

10

At a time when the Internet has opened the door for so many artists to offer their recordings to so many people, any discussion about income from recording and the involvement of record labels needs a brief preface. Artists must have recordings and live performance of the music from the recordings as the foundation of their career. Music drives every aspect of artists' earning potential for themselves and for the manager who oversees their career. Artists on private labels, indie labels, and major labels can all have success with their creative efforts. But the difference between them often is the scale of their success when success is measured in earnings, and earning potential can determine whether a professional manager could or should become involved in their career.

The role of the label continues to change. The strength of major labels has traditionally been their ability to market recordings and distribute them through the wholesale and retail markets. Labels have been full-service marketing and distribution companies for recordings, but the dramatic decline in the sale of recorded music has put pressure on them to pare their staffs and reduce the amount of marketing and radio promotion they provide. As a result, many management firms have begun to employ marketing and promotion talent themselves. But the bottom line for now is this: record labels are in the business of selling recorded music, and artist managers are in the business of developing sustaining careers in commercial music.

Until the 1950s, recording artists weren't paid royalties for the sale of recordings. They were paid in a similar manner to session musicians, meaning they were paid at the prevailing rate for a vocalist who sang at a recording session, and that was the end of the compensation. Those artists who negotiated for a royalty for recording music "only received a penny amount per sale rather than a percentage of a sales price as they do today" (Butler, 2004). For today's artists who own their own label and market their own recordings, income and earnings from the sale of recordings are immediate. In most instances, however, aggregate earnings from them can be far lower than for those artists who are signed to recording contracts with bigger independent or major labels.

The artist manager's relationship with a record label varies with their management style. Managers who have had success directing an artist's career tend to become good partners with a label, being actively involved in planning and discussing what they can bring to support an album project and how they can assist in its promotion. The managers' past successes tend to make them compatible partners with a label. However, there are some managers who bring nothing to the table, are slow to respond to requests of the label, and find themselves outside the planning process.

Our look at income from recordings begins with a contract with a large independent or major label and what it can mean to an artist and their career.

RECORDING FOR LARGE LABELS

From the artist manager's perspective, the record label represents an important career-marketing machine for the artist as well as an indirect income generator for other aspects of the artist's career, both as a writer and performer. Record labels continue to have marketing and product distribution as their strengths, and the recording contract can play a pivotal role in career advancement. Knowing that this will require the formality of a contract, the manager should seek an experienced and active entertainment attorney to handle the negotiations with the label. Entertainment law is a specialized area, with new contract provisions occurring as regularly as the music business changes—almost daily. An experienced attorney knows what he or she can get as concessions from the company and also what contract provisions might be open for negotiation for a better deal for the artist. The manager should let the attorney be the negotiator, but should also attend all meetings where negotiations are taking place. Because business negotiations can become contentious, it is better for the attorney to take the hard-line role with the label rather than the manager, as the manager will be conducting business with the label when the contract has been completed.

When you cut away all of the fine print of any contract, it is simply an exchange of promises. With a recording contract, the basic promises exchanged are these: the

Artist's income

An artist earns most of their income from the record company in the form of royalties that are due, according to the recording contract, each time a copy of the recording is sold at retail. Royalties are a percentage of the price of a recording paid to the artist, which operate very much like a salesman's commission.

The "price" against which the artist's royalty payments are made is defined in a couple of ways. Many new recording contracts specify that the royalties paid to an artist are a percentage of the wholesale price of the recording (*published price to dealers*, or *PPD*), meaning they are paid a percentage of the price that the record company charges its dealers who distribute their recordings (Passman, 2009). Simplifying the math, if an artist is paid a 10% royalty for each album sold through iTunes and the wholesale price is $7, then the artist earns 70¢ in royalties each time a recording is sold. Another way royalty payments are determined is found in many older contracts that still contain language saying that royalties are paid as a percentage of the suggested retail price minus a 25% charge for packaging. There are actually more calculations to determine royalty earnings, but this shows the basic concept of payment by the label to the artist.

Royalty rates for new artists can range from 12% to 14% for each recording sold; more established artists can earn up to 20%, and major recording artists might exceed 20% (Hull, 2004). Most contracts also include a provision that allows the royalty rate to increase as a recording reaches sales plateaus such as a half million or million units.

Here is an example of how an album that sells a half million units would theoretically generate income for the recording artist. Table 10.1 shows how the royalty base for the artist is determined for CDs and digital albums. In this example, the artist has a base royalty rate of 13% from which the producer's royalty is deducted.

Table 10.1 Royalty Basis

	CD	Digital
Retail List Price	$13.98	$9.99
Minus container charge	(3.50)	-
Royalty Base	10.48	$9.99
Basic Royalty Rate	13%	13%
Minus Producer's Royalty Rate	–3%	–3%
Net Album Royalty Rate for Artist	10%	10%
Penny Rate (Net album rate × Royalty Base)	$1.048	$0.999

(Owens, 2010)

Table 10.2 shows the result of applying the artist's penny rate (royalty rate) to the number of units that were shipped.

Table 10.3 shows the total earnings from both CD and digital sales but also lists those items for which the label typically pays but are required to be paid back, or recouped, by the label. We discuss this later in the chapter. It's also important to note that the units the calculations are based on are on "shipped" albums, meaning that stores will be permitted to return any unsold copies, which will also affect the ultimate amount of royalty payments.

Throughout the life of the recording, the label also collects royalty earnings for the artist based on the shipment of the recordings into the marketplace. As the shipments are made, the label sets aside the artist's royalties for each and holds them in a reserve account. As the royalties accrue to the artist, costs that are recoupable to the record company are paid from that account. This simply means that the label applies earned royalties to the money the artist owes back to the label for creating

Table 10.2 Royalty Earnings

	CD	Digital
Total Units Shipped	450,000	50,000
Less Free Goods (20%)	(90,000)	0
Royalty Bearing Units	360,000	50,000
Penny Rate (× Royalty Bearing Units)	$1.048	$0.999
Royalty Earnings	$377,280	$49,950

(Owens, 2010)

Table 10.3 Net Royalty Earnings

Total Royalty Earnings	$427,230
Less Recording Costs	(175,000)
Less Video Costs (50% of $170,000)	(85,000)
Less Independent Promotion (50% of $100,000)	(50,000)
Less Artist Advance and Tour Support	(75,000)
Net Royalty Earnings	**$42,230**

(Owens, 2010)

the recording. It becomes easy to see why artists may not receive a royalty check from the label for perhaps two years after the first recording was released, if ever.

For the established artist, income arrives quickly to the label, costs are recaptured quickly, and the label is able to distribute royalties much earlier than for the first album released.

The timing of royalty income is important to the artist manager because one of the typical responsibilities they have is to manage a budget for the artist. Labels distribute royalty checks for sales of recordings three quarters (one quarter equals three months) after the release of the recording. Remember that the artist must recoup certain recording and other expenses to the label before being entitled to royalty payments, so the timing of those earnings may take considerably longer than nine months to appear. The label may be able to give the manager an estimate of when and how much the royalties will be when they are distributed.

New and established artists must deal with another delay of royalty distribution, which is defined by label accountants as "the reserve for returns." The artist's royalty account is based on shipments of recordings minus the royalty value of the albums that are returned unsold. In an effort to hedge the overpayment of royalties to artists, the label usually withholds about 20% (but as much as 50% for new artists) of royalties that are due to the artist in a separate account that is referred to as the reserve account for returns. When the album has completed its life cycle, distributors return the unsold recordings to the label. The label then deducts the value of artist royalties for the returned recordings from the reserve account for returns, and then pays the balance (if there is one) to the artist (Macy, 2006).

How does an artist manager determine whether the royalty payments to the artist are accurate? Most recording contracts allow the artist a two-year window within which to request an audit to confirm that payments are accurate. However, California law extends that to three years, and it has given more latitude on the choice of the auditor who is permitted to examine the label's accounting records.

THE ROLE OF THE PRODUCER

Another element that has an impact on the earnings of the artist is the cost of paying a producer to create the recording. The producer is the individual who assembles all of the necessary elements to take into the recording studio to create the recording. This ranges from reserving a studio, to finding musicians, to helping choose songs for the artist to record. Because the artist pays for the recording, the responsibility of paying the producer is the artist's, and this is done through the artist's royalty earnings.

Also, an artist signed to a record label with an "all-in" contract means that the specific royalty rate being paid to the artist includes the royalty payment to the producer of the recording. For example, the artist who earns a 12% royalty may be required to pay 4% or 5% of that royalty rate to the producer, depending on the rate the producer negotiated, reducing the artist's earnings to 7% or 8%. Many producers require an advance paid by the record company followed by earnings from royalties. Because the record label is fronting the money to create the recording, their position is that of a key stakeholder in its success, so the company will have considerable input into who is chosen to produce the recording. The artist and producer certainly must have compatible creative chemistry, but the label will want to have the right to overrule a decision by a new artist on who the producer will be.

The choice of the producer should be based on who can create commercial music using the talents of the artist that will result in a recording that connects with music consumers. The label and the artist look for a producer who has a good or a developing reputation, and one who is a good creative "fit" with the artist. Labels also seek producers who have their own marquee value to terrestrial radio programmers as well as with the consumer. The success a producer has with other artists can help build the reputation and market the music of a new artist.

OTHER EXPENSES CHARGED TO THE ARTIST

The label has promised to market and exploit the recordings that the artist creates. However, the marketing they are willing to pay for has limits. When these limits are exceeded, the excess is charged back to the artist with their permission or that of the manager, and the artist must pay for it with royalties earned from the sale of the recording. Here are some of the charges the manager can expect to see on the artist's statement of account:

- A portion of the video will be charged to the artist. Recording contracts often specify that the label will pay half of the cost of a video up to a total of $100,000; the artist will pay for all amounts that exceed that. Competitive country videos can be created for less than $100,000, but videos for BET and MTV2 can easily exceed $250,000 dollars.
- If independent radio promotion people are hired—those not directly employed by the record label—to attain radio airplay, the artist may be required to pay half or all of this additional expense.
- If independent publicists are hired (those not under contract with nor on the regular payroll of the record company), they will be paid entirely by the artist.
- If the record company pays for any aspect of live performances, the entire amount becomes recoupable. This would include costumes, hairdressers, makeup assistants, and other stylists—plus any transportation or equipment required for a performance.
- The artist will also be required to pay all costs of creating multimedia configurations of the recorded performance for current and future technologies.
- Any tour support money provided by the label will be recoupable by the label.
- An array of other things might be charged back to the artist. For example, if the artist wants an upgraded jewel case for their CD, they will be charged the

difference; if the artist wants a full-color jewel case booklet rather than one with black and white interior pages, they will be charged an additional amount.

Any charges to the artist that are beyond the specifics in the recording contract require the approval of the artist manager before the artist becomes responsible for the expense.

Deductions from the artist's earnings also include a reduced royalty rate for recordings that are sold at various discounted price points, at military installations, and in foreign markets. Certain taxes are also deducted from the artist's earnings.

THINGS FOR WHICH THE LABEL CUSTOMARILY PAYS

The manager will find that the contract contains some things that a record label typically provides as part of its agreement with the artist. Labels pay for the manufacturing of the physical product after it is mixed and mastered.

Promotion costs are paid by the label. In this instance, label "promotion" is more like lobbying. Most labels employ promotion people whose responsibility is to influence radio programmers to choose the label's recordings to be played on the radio. The personnel costs for this service are borne by the label. However, if outside independent promoters are hired, some or all of this additional cost will accrue to the artist. Label promotion also includes those at the label who are trying to influence programmers of video channels to choose the artist's work, and those who are working with Internet websites to place the artist's music videos.

The label also pays for marketing costs, which includes advertising. The label makes advertising decisions based on the most effective way to place a paid message in front of a consumer within the target market of the artist. There is a wide array of media to carry the message, and the label will choose the most efficient way to use the available advertising budget to reach the consumer. Other marketing costs can include product design, wholesale and retail distribution, shipping, publicity, tour support, and a portion of the video production. Labels will view some of these as products of their creative services department, but they all serve to market the recording.

It becomes the responsibility of the artist manager to ensure that the artist is getting the best opportunity for success with their recorded music project. A prominent record promotion executive with a major label said, "All artists are not created equally at a label," meaning that artists have similar contracts with the label, but can be treated quite differently in the way company resources are used to promote the artist's recordings. A manager needs to have a good relationship with the promotion people at the label and to be actively and continually aware of the amount of effort the label is making to promote recordings—in part because the cash cows will always get what they need and the new artists will get what's left. This becomes one of those times when the artist manager must be a very aggressive advocate for the artist to be sure the label gives the artist a reasonable amount of the resources needed to make the recording a success. It is logical to assume that labels would want to commit resources to make all of their recordings successful, especially those of artists with multiple rights recording contracts. However, many radio promotion people are paid based on the chart position achieved by a recording, and/or retail sales of a recording, so they will put their energies where it has the most impact on their personal bonus structure.

CURRENT TRENDS IN CONTRACTS FOR RECORDING ARTISTS

Trends in recording artist contracts being offered by major labels and large independent labels contain numerous concessions that contracts have traditionally not included. Labels are seeking ways to reduce their financial risk and increase their earnings by linking to income streams that were previously reserved only for the artist. These new contracts are multiple rights agreements—sometimes referred to as "360 deals"—and likely include the following additional provisions, which make the label a partner with the artist by sharing nontraditional income:

- The label will require ownership of the official website of the artist, including the right to sell advertising and the artist's merchandise on the site. And although the label will own the website, costs to create it will be charged back to the artist and will be recoupable. If the artist is able to share the income generated by the website, the manager will be required to negotiate that as part of the recording contract.
- Income earned by the artist from touring and merchandise sales must be shared with the label. The amount of this percentage varies, depending on the label, and is an important negotiating point for the manager and the artist's attorney.
- The newest contracts give the label a license to use artwork created for the album for other products.
- The creation of ringtones, voice tones, and ringbacks for the label's promotional use are contract requirements. The new artist will also be expected to agree to the use of their image as computer wallpaper. The promotional use is not necessarily linked to an artist's album project, and because it is used for promotional purposes there will be no royalties paid.
- Labels seek to sell more music electronically and are interested in increasing the number of outlets that will sell the artist's recordings. However, as the manager will find in the contract, revenue from electronic sales of recordings is significantly lower than for the standard physical CD. As the label sells more albums online, the overall units-sold royalty income to the artist will decline.
- Contracts being offered by labels continue to include a required deduction for packaging and they create a reserve for returns for sales of singles and albums online. Online sales have no packaging and music sold online is not returnable, but expect this provision to be included in the contract.

- When an artist's recording is licensed for a use for something other than a sound recording, perhaps for use in a performance DVD, the artist can expect to earn 12% of the net earnings produced by the license. Net earnings means the income after all expenses are paid. Previous contracts offered to artists set the net income sharing with the artist at 50%.
- The manager should also be alert to the number of recoupable expenses to the artist, and to watch for an increase in the percentages of recoupment to the label of things like the cost of a video. For example, independent promotion costs had been shared equally with the label in the past, but labels now shift a greater percentage of that to the artist in the newest contracts.[1]

Most 360 deals take a percentage of the artist's income in some or all the areas of touring, merchandise sales, song publishing, and sponsorships. The manager can make a significant difference in the amount and sources of income that become part of an artist's recording contract.

An example of how a 360 contract can work, Atlantic Records offered a recording contract to an artist with these deal points:

- The label will give the artist a traditional advance to induce them to sign the contract.
- When the artist releases his or her first album, the label will have an option to take 30% of net income from touring for a payment of $200,000.
- The label will give the artist 30% of the earnings from the album, which is twice the amount of profit sharing most labels permit.
- The label will have the privilege of approving the artist's tour schedule.
- The label will be permitted to approve who the tour manager is and who the merchandise sellers are, and they can set the salaries each makes.[2]

As labels seek to find ways to improve their profitability, contracts offered to new artists will continue to include creative ways for labels to participate in the earnings of the artist. However, as artists and their managers track the value of new media developments, more income opportunities become available and they are now demanding more income for the use of their talents. For example, artists are seeking higher fees for licensing music videos in which they are featured. As *Billboard* notes, there are more wireless and online outlets that feature video as website content than ever before, and artists want to participate in the revenue that is generated (Bruno, 2005). Google, Yahoo!, YouTube, MySpace, AOL, and others generate massive amounts of Internet traffic to their websites, and record labels have found ways to profit from it and artists seek their share.[3]

---

[1]Milom (2006).

[2]Leeds (2007).

[3]The author extends an expression of gratitude to veteran entertainment attorney Mike Milom of Bass, Berry & Sims, PLC, and to *Music Row* publisher David Ross for permission to use some of the information in the previous section.

A CHANGING MODEL FOR MAJOR LABELS?

As labels seek ways to cope with the decreasing sale of recordings, the emerging model is one in which new artists are signed to label contracts as "brands," but will continue to be managed by their own independent representative, their artist manager. Among the concerns by labels in taking the complete management of an artist's career is the fiduciary responsibility managers have with their artists. A fiduciary relationship is defined as follows:

> *Confidence placed in someone else regarding a transaction or one's general affairs or business. The relationship does not need to be formally or legally established but can be based upon personal or moral responsibility due to a fiduciary's superior knowledge and training as compared to the one whose affairs are being handled.*

(Legal-Explanations.com, 2010)

Although Jay-Z, Madonna, and others have signed with Live Nation to manage all aspects of their careers, many labels have concern about the inherent conflict of interest created by a business relationship that includes a fiduciary responsibility like that of an artist manager. Most large labels seem content to participate in the earnings of their new artists but wish to leave the management to someone else.

ARTISTS WHO RECORD FOR INDEPENDENT LABELS

Independent record labels often serve musical niches that the majors do not seek, but they can also give developing mainstream artists a place to prove their commercial viability before the manager presents them for a recording deal or a partnership with a major label. When an artist demonstrates the ability to connect artistically with paying customers, the manager has a very strong position from which to negotiate a contract with a major label. The contract with the label can be for the leasing of a master recording to the major label—either an album or a single—or for a full recording contract with the label for the development of a new album project.

From a sales perspective, few independent labels have the resources that permit them to compete for consumers at the level of the major labels. As a result, they approach their business very differently, and the artist manager approaches his or her profession on behalf of the independent artist in a different way, too.

Unless they are very large companies, independent labels typically cannot afford a staff to promote a recording to radio. Nor can they afford advertising, videos, video promotion, and price and positioning at retail. They rarely have money for tour support. Compared to the 500,000 units sold for a moderately successful major label project, a successful recorded music project for an independent label sells in the range of 100,000 to 250,000 units. On a regional basis and with a close eye kept on budgets, an independent label can create income for itself and its artist—always remembering that it cannot and should not try to compete on a national level with large labels unless or until its regional success is ready for national exposure.

The manager by necessity is very involved in the independent album project, which is key to generating income from the sale of recordings. Much of the "marketing" for an independent recording involves coordination by the manager of the artist's activities to promote the release of the project, and the resulting publicity and performance dates. Some independent labels help support the project with their staff publicists and with street teams, e-teams, and web promotion. Most of the sales of recordings are the result of live performances, with the sale occurring either at the venue or at local stores on days surrounding a performance (Hutchison, 2005, 310). The growth and potential of online promotion and sale of independent music is explored in depth in Chapter 7, where we looked at the manager's work for the artist using new media.

It is difficult to appreciate the amount of effort required by the artist manager on behalf of an artist releasing a recording on an independent label without looking at an example. To support a very active touring schedule, Grammy winners and platinum-selling artists Take 6 released an album, *Feels Good*, on an independent label, which they own. Industry veteran and artist manager Chris Palmer spent months planning for the release of the recording and created an immensely active schedule of public appearances and meetings with the media to promote it. He was managing in an environment that had a very limited budget, a small management staff of two (including himself), and a few contract team members for Internet and promotional support. Periodically, he provided members of the artist's team with updates of the activities he was coordinating in his "Marketing Report." His report covering a 60-day period shows the enormous amount of planning and continuous follow-up and coordination necessary to execute a marketing plan for an independent label. The Marketing Report is reprinted with special permission in Appendix A. Some major labels and large independents handle these activities on behalf of their signed artists but very often, as in the case of Chris Palmer, it falls to the artist manager.

IT'S BUSINESS

In all of the dealings that the artist manager has with the record company, it is important to remember that the label is in the business of selling recordings. The key word here is "business." And when the artist is signed to a label, they become an asset, much as any other item of value listed in the company's annual report. Assets are managed to the financial benefit of the company until they no longer have value. In the case of artists, when their recordings are no longer selling, they are no longer an asset, and the company ends the business relationship. The joy of the shiny new recording contract and promises of stardom—will eventually become a visit by a label representative from top management with the message that their business with your artist is over. Sometimes the label will deliver the news personally to the artist, but it often becomes the regrettable duty of the artist manager to be the messenger.

Through it all, always remember that this is business, and that selling recordings and associated products drive the bottom line for the record company.

THE ROLE OF RADIO IN THE RECORDING ARTIST'S INCOME

Many of the traditional forms of promoting the artist are relatively straightforward. For example, when an artist appears in a town for a live performance, there is publicity through the media, regional fans are contacted via email, music is played on radio, advertising is purchased, tickets are sold, the performance sells seats, and the artist sells recordings and merchandise as a result.

Traditional commercial radio has helped promote the sales of music through airplay, but shrinking listening audiences continue to reduce the amount of influence airplay has on consumers. For example, listeners between 12 and 24 years old have cut their radio listening in half, and nearly half now learn about new music from their friends (Webster, 2010). It is becoming increasingly difficult to determine the measure of influence radio has in music sales but it continues to be a necessity for the promotion of an artist's recordings. Although the role radio has played in the promotion of an artist's career may seem obvious, the way radio uses that relationship is not so obvious. In this section, we look at traditional terrestrial radio, what the priorities are for radio and its owners, and how the artist manager can use this information for the benefit of the artist's career.

The business of terrestrial radio

It is important to understand what business traditional radio is in, and what business it is not in. Radio is not in the business of selling recordings, promoting careers of artists, selling concert tickets, nor working to assure a healthy music business. With the exception of promoting a concert and selling tickets, none of the activity surrounding an artist contributes to the profitability of the radio station and is not important to its business other than for its own promotional value. What is important to the radio station is its number of listeners. Veteran radio programmer Lee Logan once said that the business of radio is to build an audience to lease to advertisers, and that is as succinct as one can put it.

Let's look at the first part of that statement. Radio builds audiences. The purpose of the entertainment you enjoy on radio is to attract your attention and keep you tuned in. Even though radio audience measurement company Arbitron continues to show that talk radio is the most popular radio format in the United States, over 80% of the remaining radio stations program some form of music to entertain its audience (Arbitron, 2009). In order to build a particular station's audience, the program director—who makes all decisions about everything that goes out over the air—seeks music and on-air talent that will keep its current audience and attract new listeners.

Now, to the second part of Lee Logan's statement. Advertising rates are based exclusively on the number of people who listen to a radio station, so the larger the audience that the programmer can build, the more the station can charge for its advertising, and the more money the business will make. For example, in Los Angeles, Arbitron audience ratings for station KZLA continued to decline from 2.1% share of the city's radio audience to a rating of 1.7%. The result was that the station replaced personnel and changed its format from country to a contemporary hit format. Why? They switched formats because a change of 1/10 of 1% of the audience share in Los Angeles is worth 1 million dollars of advertising revenue per year for the station owners. KZLA ownership felt they would draw a larger audience share with a different programming format. The same principle applies to commercial radio stations everywhere, although the financial impact depends upon the audience size of the radio market.

The point of this is that the artist manager must understand what is important to commercial radio. In order to get anything done through someone else, you must understand what they need. In the case of radio, they need growth in their audience share. As a result, programmers are very careful to play music that will keep current listeners and attract new ones, and they will program their stations to be predictable by meeting the listener's expectations when they tune into the station. If the manager's artist is too different from those whose music is being played on the station, this conservative nature of most programmers means they will not schedule the artist's music for airplay.

Most music will never be heard by a commercial radio programmer or music director without aggressive one-on-one promotion by someone skilled and experienced in getting it done. If the artist is signed to a major label or a large independent, or is managed by a company with a radio promotion budget, the companies will have a staff that promotes recordings to radio. Programmers will know that there is a significant promotional effort behind a recording, and it will be apparent that the artist's recording is intended for mass commercial marketing. It is easier to get the consideration of a programmer under these circumstances than if the artist is new and is on an independent label with no affiliation with a major.

This is not to say that the new artist the manager just signed will never find their way to radio. A continuing classroom exercise by the author examines the *Billboard* Top 200 sales chart and finds that for most business quarters except the fourth, 20% to 30% of the artists who are in the listing of the top 50 selling albums are relatively new artists. Considering the number of veteran sellers of hit recordings who crowd the top of the sales and airplay charts, these observations show that there is opportunity for a new artist. We liberally define a new artist for this purpose as one who has been active with his or her first commercially viable single and album over the last year or 18 months.

The business of satellite and online radio

Satellite radio is a subscription service in Canada and the United States that gives users a vast array of entertainment on special radio receivers designed to deliver their programming. The service offers scores of music channels, many of which have deep playlists that include new music. There is opportunity for new artists to find a place for their music on satellite radio, but it requires an experienced radio promotion person to reach the gatekeepers.

Online radio, in its range of offerings from web feeds from standalone stations to sites like Pandora, also provides opportunities for artists to showcase their music. The advantage of online music stations and subscription services like Napster is that the listeners have the opportunity to immediately purchase and download the music they are experiencing.

The charts

There are two kinds of "charts" that the manager should understand—sales charts and airplay charts. *Billboard* publishes its Top 200 each week, which is a ranking of the top-selling album recordings in the United States, but does not include sales data to support the ranking. The publication includes the weekly chart that ranks the position of the recording based on sales, but it does not show actual sales data. The sales data is proprietary information that is available only to those companies that subscribe to the data services of SoundScan. Each time a recording is sold in the United States at a retail store or online, SoundScan captures the information and reports it to its central data assembly location. The seven days of sales that make up the Top 200 chart end at Sunday at midnight EST, and the sales data is reported early the following Wednesday morning to SoundScan clients. As a point of reference, there is no company anywhere other than SoundScan that compiles verifiable and actual sales data of recordings.

The other kind of chart the manager should understand is an airplay chart. Whereas the Top 200 shows the sales of albums, airplay charts show the number of times a single has been played on commercial radio. A number one song is the *single* that receives the highest amount of radio *airplay* for a seven-day period. (A number one *album* is the one that *sells* the most within a seven-day period.) *Billboard* magazine publishes airplay charts each week in its magazine, which is a guide showing how often songs are being played. Similar airplay information is compiled by another company called MediaBase, which is published each Tuesday in *USA Today* and through other publications and websites.

Airplay charts are convenient guides to radio programmers so that they know what songs their counterparts in other cities are choosing to play most often. Radio programmers have wide and varied duties at the radio station and making decisions about which music to play, and which music to stop playing is just a part of their responsibilities. To the programmer, the airplay chart is used as a reference, or barometer, of how songs are performing elsewhere. The research, experience, and understanding a programmer has of the local radio market are perhaps the most important criteria in the music selection decisions.

College radio

College radio in recent years has picked up considerable competition from the Internet as a source for filtering new and interesting music. However, it still has enough reach and impact that the manager of an artist should not overlook its influence. In the 1980s, college radio was the source for music that was overlooked by commercial radio because it was generally viewed as being outside the mainstream. The alternative music format began at college radio and now has its own commercial format following (Calderone, 2005). Because college radio is noncommercial, it has limited resources to market and promote itself. The traditional programming of college radio is to offer small blocks of time during a broadcast day to introduce audiences to a broad array of music that defines the adjective "eclectic."

The size of college radio audiences is often much smaller than the commercial counterpart. In a midsize southeastern U.S. city, the suburban university's college radio station has 2,000 unique listeners each week. Its competition at a nearby urban university has a weekly audience that is nearly ten times as large. By commercial radio standards, this is still a relatively small listener base, but it can be an important promotional avenue for the artist building a fan base and who is on tour playing at venues appealing to college students.

SPONSORSHIPS, ENDORSEMENTS, TELEVISION, MOTION PICTURES

Among the things that an active and visible recording career can provide is access to other ways to earn income and to exploit the talents of the artist. Sponsorships and endorsements are offered to an artist when the target market of a brand are the same as the artist's, or the target market of the artist is one that the brand seeks to have as its customer. The manager and sometimes the agent will be able to negotiate a sponsorship or product endorsement on behalf of the artist.

Endorsements are made through those who control the commercial messages for the specific product, such as product brand managers or the person who handles the product account at an advertising agency. It will take a little research on the manager's part to determine who that person is, but they will be the one to decide whether the image of a product benefits by associating with an artist. And this applies to local or regional products or services, not just national brands.

Carolyn Ballen from the Indie Music Forum offers some ideas that can be presented to a possible sponsor about the benefits that sponsoring an artist could be, including:

- An endorsement from the artist in the sponsor's advertising
- The sponsor has access to a large following of people within the sponsor's target market
- A logo presence via a banner or product demonstrations on the grounds of performances
- An ad in the artist's CD
- Mentions about the product from the stage
- A logo on all printed materials, including advertising, and logos with links on the artist's website

- Promotion of the brand in all fan club emails and other consumer communications[4]

There may be a temptation for an artist-songwriter to include a sponsor mention in his or her song lyrics, but radio programmers are sensitive to the use of this Trojan Horse–style commercial promotion and will likely not program the song.

Sponsorships can amount to a few thousand dollars for a local or regional touring band, but easily goes into the millions for major artists. Rapper 50 Cent agreed to allow vitaminwater® to sponsor him, and the resulting income exceeded $100 million.

Television and motion pictures can add an additional dimension to the career and income of an artist, which is often done with the help of a full service booking agency such as Creative Artists Agency or William Morris Endeavor Entertainment. These organizations have the in-house resources to represent the artist in both their music as well as for any acting opportunities that may be appropriate for the artist. Movies featuring acting roles by music artists Justin Timberlake, Beyoncé Knowles, Ludacris, Queen Latifah, Tim McGraw, and many others allow their talents to grow beyond the stage of live musical performance. It is incumbent on the artist manager to encourage artists to develop in creative areas that can take advantage of the brand they have developed and draw from its value in other areas of entertainment.

There are also opportunities for artists' recordings based on the successful model Disney developed, with its soundtracks and artist spin-offs from the television shows *High School Musical* and *Hannah Montana*. Both became multiplatinum-selling recordings with little radio support except through Radio Disney.

References

Arbitron. (2009). *Format Trends*. Retrieved from http://arbitron.com/home/content.stm.

Ballen, C. (2007). *Looking for Sponsorship?* Retrieved from Starpolish.com.

Bruno, A. (2005, October 29). *Video Booms Online—But for Whom?* Billboard.com. http://www.allbusiness.com/retail-trade/miscellaneous-retail-retail-stores-not/4555125-1.html.

Butler, S. (2004, May 7). The Publisher's Place: Clause and Effect. *Billboard.*

Calderone, T. (2005). College Radio Grows Up; Nine Inch Nails Returns. *New York Times.*

Gordon, E. (2005, June 17). *News and Notes with Ed Gordon*. National Public Radio.

Hull, G. P. (2004). *The Recording Industry*. Routledge, NY.

Hutchison, T., Amy Macy, & Paul Allen. (2009). *Record Label Marketing*. Burlington, MA: Focal Press.

Leeds, J. (2006, December 11). Squeezing Money from the Music. *New York Times.*

Leeds, J. (2007). The New Deal: Band as Brand. *New York Times.* Retrieved from http://www.nytimes.com/207/11/11/arts/music/11leed.html.

---

[4]Ballen (2007).

Legal-Explanations.com, (2010). http://www.legal-explanations.com/definitions/fiduciary-relationship.htm.

Legrand, E. (2004, August 14). Global Music: European Indies Rise Up. *Billboard*.

Macy, A. (2006). Personal conversation.

Milom, M. (2006). *The Impact of New Business Models on Artists*. Music Row Publications, Inc. Nashville, TN.

Morris, M. (2004, September 25). Taking Issue: Good News for Artists. *Billboard*.

Owens, C. (2010, February/March). Artist Royalties from Gold Albums. *Music Row*, 30.

Palmer, C. (2006). Personal conversation.

Passman, D. (2009). *All You Need to Know About the Music Business*. New York: Free Press.

Satzman, D. (2003, February 24). Radio Ratings Service Again Under Fire for Methodology. *Los Angeles Business Journal*.

Webster, T. (2010). *The American Youth Study, Edison Research*. http://www.edisonresearch.com/home/archives/2010/09/the_american_youth_study_2010_part_one_radios_future.php.

Conducting business for the artist

11

Part of the business of managing the career of another includes the efficient use of time and embracing ethics in the ways that others would expect—especially because you are acting on their behalf. This chapter looks at time management and the ethics that go into your management of that time.

PRESENTING THE ARTIST FOR A RECORDING CONTRACT: AN EXERCISE IN TIME MANAGEMENT

Among the most valued assets an artist manager has is time. How and where managers use their time on behalf of their artists dictates how efficient they are in the other areas of artist management. Managing time means that an artist manager must develop the ability to manage meetings—large and small—even if the meetings seem to be in the control of others.

The manager schedules meetings for a variety of reasons, but none is more important than the meeting requested with a label to present an artist for a recording contract—not because it creates the largest income stream, but because the marketing and promotional efforts by a label help drive interest by fans in touring. Given the weight of this meeting, we'll use it here to carefully create a template that will have a wide variety of applications to the work of the artist manager.

The meeting has been set for a Tuesday morning at the offices of the record label. The manager has asked for 15 minutes with the key staff of the label, and the label president if he or she is available.

If you think you can never be effective in a meeting like this, read through this section and you'll learn that it may not be as difficult or impossible as you think.

Going for the record deal

Even for the veteran manager, planning and executing a meeting like this can create anxiety and stress. Knuckles get tight, palms become moist, feet get cold, and butterflies invade the stomach. If the manager embraces the true importance of this meeting about the future of the artist, some of these physiological symptoms will appear, but they are completely normal reactions. How does a manager deal with this? The surest way to handle this normal reaction is to plan, plan, plan for the meeting. The success of this meeting depends entirely on the amount of planning the manager does to prepare for it.

The conference room environment can be very intimidating. It typically has a big table, big windows, fancy carpet and chairs, and racks of AV equipment that an MIT graduate engineer would have difficulty operating. Meeting rooms often have more chairs than are necessary and the general massive nature of the room can seem menacing. Adding to the threatening nature of conference rooms is that the meeting attendees are on "their" turf. However, even the most reluctant manager can overcome much—if not most—of the anxiety of a planned meeting simply by following these steps, which are specifically designed to put the manager in control of much of the meeting, instill confidence, and make the manager an effective advocate for the artist.

Know the purpose of the meeting and do the homework

The objective of the relationship the artist has with the label is to sell recordings and related products. The manager who presents an artist to a label for a record deal is asking them to risk hundreds of thousands or even millions of dollars and to sign their artist ahead of any others being considered. Likewise, a small independent label is being asked to risk their very limited resources to support a recording artist's commercial music project. In both cases, the risk is significant to the individual companies.

Although some business meetings are intended to inform people or perhaps to build relationships, the manager's planning must be focused on the idea that this particular meeting is a sales meeting: the objective is to convince those in the room that the manager's artist has a place on the label's roster and can make money for the label. An important thing to remember about a meeting like this is that it is not about the artist—it is not about the manager. This meeting is completely about the record label and what *they* need. This basic understanding about the focus of the meeting—that it is totally about the label—must drive the manager's planning for and execution of the meeting.

Whenever one presents in an environment like this, it is important to know what those in the room need. After all, the only reason people would attend a meeting like this is if they think there is something in it for them or their company. The label is in business to make money for the label ownership through the talent and artistry of their signed recording artists. So it is important to know what they need before attending the meeting.

Among the things about artists that the label will consider important are:

- The music they write and perform
- The image they present on stage and for the camera
- Their self-assuredness and confidence in who they are
- Their experience and seasoning as an artist
- Their uniqueness and contemporary appeal
- How strong the artist manager is
- Whether the artist has a publishing contract with another company
- How much touring the artist has done and how much is planned
- Sponsorship and merchandise deals the artist has
- Their experience selling tickets and recordings and associated successes
- Internet traffic to the artist's social networking sites
- The maturity of their overall Internet presence

These points are important to the label because they address the commercial viability of the artist by demonstrating their preparedness for a career in commercial music. They also disclose opportunities the label has in minimizing their risk by seeking partnerships with the artist by sharing some of the existing and future revenue sources. New artist recording contracts, as we know, usually include revenue sharing in the form of multiple rights 360 contracts. Knowing these points of interest to the label should be part of the meeting planning by the manager.

Likewise, the manager should be aware of other aspects of the label that identify needs that are not specifically related to the artist being presented for a recording contract. Among those things are:

- How many similar artists the label currently has under contract
- Whether the label has budgeted to add an artist to their label
- How much of the promotional work the manager will be expected to handle
- Whether the label has the staffing necessary to support another artist
- How risk-averse the label's reputation is
- The number of acts that are on the label's roster
- Whether the roster has been recently cut, making way for another act

All of these points are among those a manager should consider in planning to address the needs of the label in preparation for the meeting.

The manager's research about the first label that will be approached should be very complete. The manager should know what the recent history of album and single sales has been for the label, and what the financial condition of the label is. Finding someone who has access to SoundScan will give you this important inside information. Also, the manager should know about the politics of the label, such as whether there were any recent personnel changes or rumors of any that are pending.

Prepping for the meeting

Industry trade magazines and websites should be among the first sources an artist manager employs to acquire background information for planning the meeting with the label. Key trade magazines for this research include *Billboard*, *CMJ*, and *Pollstar*. These sources will include airplay and sales charts as well as information categories that include activities within specific labels. *Pollstar* magazine gives weekly ticket sales by geographic location and by venue. Websites with current information about labels include *http://allaccess.com* and *http://billboard.biz*. Dean Kay's free daily aggregation of music industry news is also a valuable resource noted in the reference section at the end of this chapter. All of this information will be important in giving some insight into the needs of the label for which the meeting is planned. If the manager understands the competitive and changing business environment the label faces, it will help the manager see the needs of the label in terms of the artist being presented for a recording contract.

It is helpful to know which labels are signing artists and which ones are not. In the manager's research, knowing that artists have been cut from a label indicates a possible opportunity. And simply because the meeting planning research shows certain labels are not signing new artists, the truth is that a label that discovers an artist who fits their commercial vision will consider adding that artist to their roster. The bottom line to record labels and the recording industry is the "bottom line," and no label would pass up what is viewed as a genuine opportunity for success.

Planning for results

It isn't always apparent in advance how many people will attend the meeting, but the manager should specifically plan for what will happen in the room as a result of the meeting.

First, the manager will want all attendees to *listen* to what is said. This means that the artist manager should be positioned in the room so that everyone in the room can hear what is being said. Often, the best place for one to be assured of being heard without difficulty is at the end of the table. As the manager presents, he or she will want them to *comprehend* why the meeting is being held and why it is important for them to listen. Certainly the manager will want to be able to *influence* their thinking about the artist with the goal of having them *take action* to sign the artist to a recording contract.

Knowing this is a sales meeting, the manager should develop several key points that he or she wants to make during the meeting. Actually listing the points on a notepad or in a daily planner and placing it on the table during the meeting will prevent that annoying "mind fade" when people forget things they had wanted to say. Putting these key points on paper will organize the manager and will ensure that an important point that might make the deal won't be overlooked.

The manager should organize the flow of the meeting from beginning to end both mentally and on paper. In a perfect world, the plan for the meeting will flow from an opening to the key points, to the closing, and to a call to act. Meetings of all kinds, however, never follow a script. There is always an agenda of the record company to sell recordings, there is the agenda of each department of the label, and there always is the personal agenda of everyone in the room. However, having a plan to direct the flow of the meeting will help the manager keep a focus on the objective on behalf of the artist and will help the manager get a meeting back on topic if it veers in the wrong direction and wastes the time you have reserved with the label personnel.

If the manager can know the job titles of those who will attend the meeting, it will be possible to arrange talking points specific to them. For example, if someone from A&R (Artist and Repertoire) attends the meeting, it will be important for them to know that the artist is an active songwriter and has had several songs recorded by other artists. Keeping the success of the artist in terms of the personal success of the people in the room will create a stronger connection with those whose opinions determine whether the artist gets a recording contract with the label.

Another element of planning your meeting for results is to visit the building where the meeting will take place a day or two before it is scheduled. Even if the manager is an occasional visitor to the building, traveling the route planned for the day of the meeting can assure punctuality. It is possible that highway construction or building maintenance will cause delay, and being on time for an important meeting says a lot to the people from whom the manager is seeking time for a meeting. If the manager is late for the meeting, it may be the last meeting the label staff will agree to schedule with the manager. If the artist is to be at the meeting, the manager should personally plan to meet the artist and provide transportation. Do not assume that the artist will be punctual if left to provide his or her own transportation to the meeting.

Visiting the meeting room prior to the appointment can give the manager a comfort level of being familiar with the environment in which the meeting will take place. This premeeting visit will eliminate unknowns. It is a simple matter to call the label and ask for an opportunity to visit the meeting room a few days before the meeting, saying that it will include a multimedia presentation and it is necessary to see what resources are available. Eliminating as many unknowns as possible through planning can reduce the stress and trepidation meetings like this can create for some artist managers.

Another element of planning for meetings is to be sure that the support materials are ready. The press kits should include the latest promotional creations for the artist—printed photos, a CD with three or four songs, a DVD if it captures a good performance or performances, a good bio that sells the artist, web addresses for access points to the artist such as a branded website, and a brief EPK (Electronic Press Kit) if one has been developed for the artist. Plan to take a briefcase with a dozen press kits to the meeting. Although only two or three people may be at the meeting, provide enough so they can give a copy to others at the label who will influence the decision to sign the artist.

Budget the time

Earlier, we said the amount of time reserved at the label's offices for this meeting was just 15 minutes. There are reasons for setting such a brief meeting. First, if you ask for a half an hour or even an hour, it will be difficult to set a time for the meeting. Nearly everyone has had to suffer through meetings that should have been over in 15 minutes, but instead the meeting organizers took 60. The manager who takes 15 minutes, makes the points necessary, and gets out on time is the manager who will find it is easy to get an appointment again. Perhaps this particular artist will not get the recording contract, but the manager might have another artist next month to pitch for a record deal. If the manager gains the reputation of asking for 15 minutes and taking only 15 minutes, it becomes much easier to do business with other busy people.

In a meeting where a manager presents an artist for a record deal, how does the manager allocate the time? Here's a reasonable template:

Amount of Time	Use of the Time
Two minutes	Introductory talk, light subject(s)
Ten minutes	Present main points
Three minutes	Restate main points, bring meeting to a close

To keep aware of the time, there are several strategies the manager can use. The manager can plan to be seated in a position to see the clock in the conference room; glance at someone else's wristwatch who is in the meeting; or set a wireless device to its vibrate alert as a reminder when time is nearly over. Whatever method you use to keep up with time, remember that a key to getting the next appointment will be based, in part, on how well you respect the time of others in the meeting. As a matter of courtesy, do not place a communication device on the table during the meeting: the message it sends to others is that there may be something more important than this meeting.

To the question, "What if the people in the meeting continue to ask questions and begin to run the meeting late?" remind those in the meeting that time is running short and ask if they have time to continue the meeting for a while longer so you can give them all the information they need.

Practice the meeting

If this is the first time you have conducted a meeting like this, practice it. Certainly, you can't know the flow of the meeting, but your planning will have you prepared for as much of it as you can. Having someone sit through a mock meeting with you will help you gain confidence when the real meeting happens. Ask the person to critique you by watching your gestures, body language, and mannerisms. Find someone who can muster the attitude of being brutally honest when asked for a candid evaluation. It's not time to hear what people think we want to hear; it's time to get solid input about your presentation style.

Perhaps the most important thing a manager can do during the presentation is to have eye contact with those in the room. The aim of the meeting is to convince the label that the artist is a good business fit with their company, and good eye contact tells people that you believe what you're saying.

If there are questions the manager can anticipate, practicing the answers will prevent the appearance that he or she doesn't know the artist as well as they should.

The meeting

How well the manager presents the artist as a potential asset for the record company to those in the meeting determines how productive the meeting will be. Productivity in this sense means the likelihood of a recording contract.

Assuming the artist is not attending this meeting, the manager enters the conference room confidently, smiles, and looks each in the eye as he or she shakes hands with everyone around the table. The initial smile and eye contact are important to quickly develop a friendly relationship with the group. And in this new age of business, no consideration is given to the gender of the meeting attendees when greeting others with the traditional handshake. The manager distributes a business card to everyone in the meeting, even to those he or she knows. Most will give the manager one of their business cards, which is used by the manager as a reference to names and titles of those in the meeting.

When the manager dresses for this meeting, it should be in attire that is comparable to those in the meeting. The manager wants to be viewed as a peer at this meeting and doesn't want to be over or underdressed. A quick way to determine what is appropriate is to call the receptionist at the label and ask what the label staff typically wears to meetings like this one.

Spend the first part of the meeting with a very light subject such as, "Congratulations on the new album you just released." Then let them talk, and listen for clues that might help you with your presentation. Remember that this meeting is about the label, so the key points listed on the manager's notepad are pointed toward the label and how well the artist fits into the culture and business of the label. The manager should discuss those key points, play a song from a CD or show a video of the artist, then restate the key points, bringing the meeting to a close.

Some in the meeting may have seen the artist at a showcase or at a venue where he or she performed. If the manager senses that a live performance may be necessary, consider inviting the label executives to a planned performance on a specific date or have the artist join the meeting for an acoustic performance at the meeting.

During the meeting, some of the attendees will ask for information that the manager doesn't have at the moment. When someone asks for something and the manager agrees to provide it, write it down immediately. First, doing so ensures that the manager will not forget the promise, and second, it sends the message to the requesters that their needs are important to the manager. When the manager returns to the office it is time to follow up on those promises.

A note about those in the meeting: the manager should not be offended if, during the meeting, some seem distracted or even appear arrogant. Keep in mind that many staffs at labels—both large and small—have frequent meetings where managers and others present artists for recording contracts, and this meeting may be one of several that the staff has been required to attend over a short period of days. It isn't personal.

Should the artist attend the meeting?

The primary considerations by the manager as to whether the artist should attend the meeting are: (1) whether the artist will add to the meeting, and (2) whether the artist feels comfortable in settings like this. If the manager seeks to control the meeting, it may be preferable to not have the artist attend. With the artist in the room, there will be some expectation that they will participate in the meeting and the other attendees may draw the artist into discussions that do not address the point of the meeting. However, an affable charismatic artist in the meeting room can add a dimension to the appeal of the artist that the manager cannot communicate.

Another consideration is whether the artist will be a distraction to the points of business the manager plans to make to the label executives. If the artist is a group of five people, their mere presence may constitute a continuing distraction, whereas without them in the room it becomes easier for the manager to keep the attention of the attendees on points the manager wishes to make. Under circumstances where the label asks for the artist to be part of the meeting, the manager should accommodate that request by prepping the artist.

Legendary artist manager Sharon Osbourne tells of the time in 1981 when she was presenting Ozzy Osbourne in a conference room to a group of Epic record company executives. She wanted to do something memorable and told Ozzy, "We'll take some doves and release the doves and everyone will say, 'How nice, how lovely.'" Ozzy says about his preparation for the meeting, "I had drunk a bottle of cognac and done a lot of other crap, and Sharon says, 'I'd like to go in there and make an impression and throw these doves in the air.' I says, yeah, I'll make an impression." During the meeting, Ozzy indeed released the doves from his pockets into the conference room. However, one landed on his knee and he put the head of the bird into his mouth and ripped the head off, totally shocking and sickening the Epic execs. Ozzy and Sharon were asked to leave the Epic property, but the label decided to release the album anyway. Within eight days of its release, the album, *Blizzard of Oz*, went gold (VH1, 2004). So, to the question of whether the artist should attend meetings with company executives, the answer is, "It depends."

Ending the meeting

When the meeting is coming to an end, the manager should never simply say "thanks," and leave the conference room. Rather, the manager should restate the reasons why the label should sign the artist and ask when it would be convenient to have a follow-up conversation either in person or on the telephone. In most circumstances, the apparent leader in the room will say something like, "We'll get back to you." The manager should never accept this. Rather, the manager should tell the group that he or she will get back to them in a few days to continue discussions for a recording contract. Specify a day and time when you'll call back to follow up. The important thing here is not to be put off by the label, but instead, put them on notice that you'll contact them and that the recording contract is your priority. While you are in the room, enter a date and time in your personal organizer.

Go around the room and again shake hands, say thanks, and say that you'll talk with them in a few days.

On the way out of the building, the manager must be sure to stop and thank the assistant who set the meeting that has just ended. Developing an appreciative relationship with this important assistant and gatekeeper will make setting the next meeting easier for the manager.

When the manager returns to the office, he or she should send a quick note on company letterhead thanking the key executives who were at the meeting for making time to meet, and remind them that you will be calling in a few days.

ETHICS AND PAYOLA

The music business is one of those industries that has little regulation by governments, but there are some areas of legal quicksand that a manager should avoid. We will discuss those, but first, let's take a look at ethics and what they mean to an artist manager.

Ethics

Ethics does not pertain to law. Ethics does not pertain to regulations. It does have everything to do with things like personal values of the manager, moral principles the manager uses in his or her work for an artist, and the general standard of conduct the manager uses. Each of these has been used to define ethics. But at the heart of it all, ethics in artist management means using the highest standard of conduct in all business dealings, both personally, and on behalf of artists. An often-cited definition of integrity fits well into any discussion of ethics: doing the right thing because it is the right thing to do.

Assets of managers used in the management of artists in the music business are no more important than having a strong personal reputation. A good reputation is the perception others have of another based on high ethical standards of being fair, caring about others, having integrity, keeping promises, as well as other attributes.

Whether being truthful in all business circumstances is a requirement of being "ethical" is a question for debate among philosophers, because many business deals begin with and are negotiated by using an inflation of the truth. However, being truthful with the artist *is* always required. The artist is the employer, and he or she have an expectation that the manager will always be truthful and candid. A good reputation built on ethics can make any artist manager very effective as an advocate for clients, and it is important asset to protect.

Earlier in this book, we mentioned a lawsuit brought by the Backstreet Boys against their former manager for various reasons, and it serves to underscore that the relationship between the manager and the artist is built on trust. Even though the suit did not allege any laws were broken, the dispute directly challenged the ethics of the manager in the preparation of the management contract while dealing with some very young minors.

Again, we refer the reader to the code of conduct of the Music Managers Forum in Australia in Appendix F of this book as a guide from which artist managers should model ethical conduct on behalf of those they manage.

Payola

Payola is when someone offers a radio station something of value to play a song without disclosing the payment on the air or to the station manager. The violation of the law occurs when the radio station accepts the payment and does not disclose it publicly on the air, and it also occurs when someone offers something of value to a radio station employee to secure airplay but with no intention of disclosing it to station management. The former attorney general for the state of New York, for example, found that a record label gave a radio station employee a laptop computer for the promise to play specific music (Harris, 2006) and his investigative work resulted in tens of millions of dollars in settlements with major labels and major radio groups in the United States.

Why should any of this be of interest to an artist manager? An effective artist manager must become acquainted with a dozen or more key radio programmers who decide if their artist's songs get on the radio—whether the programmers are at smaller stations in the artist's touring region or major stations. As the advocate for an artist's career, it is important that the manager builds and maintains relationships with key radio programmers. They are the gatekeepers that labels lobby to persuade to add new music to their playlists, but—as was pointed out elsewhere in this text—the manager cannot assume the label will do its job in promoting music by the manager's artists. Labels have every reason to aggressively promote the artist's latest song, but as we have learned, "all artists at labels aren't created equal," and aren't treated as if they are. As artist managers steps in to promote their artists to radio programmers, they must understand that there is a blurry line between promotion and payola. Attorneys for record labels and radio stations regularly meet with their clients to train employees about how that line is defined and to develop assurance that it is not crossed. Taking a programmer and a significant other to a nice dinner is promotion; giving plane tickets for them to join you for dinner in a resort city is illegal.

Someone summed up the payola problem leading up to Spitzer's investigation this way:

> *They're running down the field so close to the line you can't tell if they're in bounds or out of bounds, but they sure do have a lot of chalk on their shoes.*

To the artist manager, keep chalk off your shoes and get the most current advice available from an entertainment attorney so that there is a clear understanding of how to properly and legally interact with radio.

References

VH1. (2006). *Behind the Music: Ozzy Osbourne.*

Harris, A. (2006, December 29). Radio's Entercom Will Pay $4.5M to End Probe. *The Tennessean*, 3E.

The artist career plan

AN INTRODUCTION TO THE PLAN

A plan is often referred to as a roadmap to get you from where you are now to where you want to be—here's where it starts, these are the steps, and there is the goal. Just follow the plan and you'll achieve the objective. The theory works perfectly for things like baking a cake. However, the plan for the career of an artist who is newly signed with a manager certainly has a beginning, but that is where the parallel with baking a cake ends. The music business is so highly competitive and changes so quickly at the whims of pop culture and technology that the steps to achieve the goal need to flex with reality. The practical window for a career plan is no more than three years, and even in the third year it is unlikely that it will look the same after the many revisions the plan will undergo.

The career plan is far from being something that is chiseled in stone and to be strictly followed. Any plan must be flexible, and the artist career plan will require modification on a monthly basis—perhaps more often. A plan like this is a guide that is a reflection of reality, and the artist's reality will change very often the first year or two of a new career. The plan will need to be modified as circumstances change. For example, some goals will be achieved quicker than anticipated and, likewise, some will take longer than expected. There is nothing wrong with this. It is the reality that an ambitious plan designed to navigate a highly competitive industry will require continuous updating.

The career plan for each artist on a manager's client list provides an ongoing guide for both the manager and the artist. And it also gives each artist an outline of the priorities that were made in collaboration with the manager, how those priorities became goals, and how time will be used to achieve the goals. Notice the word "collaboration." It is a key concept in the creation of a career plan for an artist. The manager has a vision of how the plan should look and what the goals should be. That vision will include the general steps that are necessary to get artists where they should be given their talents and the commercial marketplace for them. However, the artist career plan is not a plan without input and agreement by the artist. Artists must see the plan as achievable before they will genuinely be willing to commit the time and energy to its success. And the manager should prod the artist to join him or her by setting the goals of the plan as high as possible without going beyond what both feel are realistic. Artists sometimes set career goals below their abilities, ensuring that they will be successful and avoiding the disappointment of failure. Know that artists will sometimes go along with setting of goals extremely high, being fully aware that they are unrealistic and they ultimately will be sabotaging their career. It becomes the role of the manager to find the point that creates a realistic career plan that the artist genuinely buys into.

Some sections of the artist career plan will include information that is very familiar to both the artist and the manager. For example, if one of the artist's talents is playing any kind of guitar ranging from four strings to twelve, it is something both are very well aware of. However, part of the reason to create a *written* plan is to allow for the possibility that someone else may need to take over management of the artist. The substitute manager may not be immediately aware of this range of talents being developed without the plan as a reference. Someone else might be called in to help manage an artist's career because the manager may become ill, the manager may have a sudden increase in his or her client list, or one artist on the client list may require immense amounts of time and someone else may need to step in temporarily to help with managing the artist. The written plan should include all details that might be obvious to the artist and the manager because they may not be obvious to another manager or an assistant.

As the artist manager begins to put together the plan for a new client, there will be the temptation to cut and paste sections from previous artist career plans, assuming that any exist. Remember that each plan is as unique as the artist is, and that six months can see important changes in the music business that the manager must be sure to incorporate into the latest artist plan.

An extensive outline of the plan is presented at the end of this chapter, In a capsule, this is the plan outline we will discuss in the following pages:

Recording artist business plan

Section I. About the artist
> Musical genre
> Biography
> Talents
> Experience
> Uniqueness of the artist

Section II. Evaluation of the artist
> Strengths and weaknesses of the artist
> Opportunities and threats
> Action points based on this evaluation

Section III. Evaluation of the manager
> Strengths and weaknesses of the manager
> Opportunities and threats
> Conflicts of interest

Section IV. Establishment of goals and timelines
> Major goals for the artist (and sample strategies and tactics to achieve them)
> Goals supporting major goals (subordinate to major goals and sample strategies and tactics to achieve them)
> Timeline for each goal

Section V. Development of a marketing plan
> State of the industry
> The target for the artist
> Detailed plans to reach the target

Section VI. Business framework
> Form of the business
> Personnel requirements
> Insurance
> Other

Section VII. The financial plan
> Personal budget for artist
> Budget for career plan

Section VIII. Exit strategy
> The artist in a mature career
> Plan by the manager to end the relationship

Because the artist career plan is collaboration by both the manager and the artist, some of the point-by-point discussion of the plan in the rest of this chapter includes considerations from both the perspectives of the manager and of the artist.

ABOUT THE ARTIST

The first section of the plan focuses on the artist. The artist is the central part of the business relationship and actually defines the artist as a branded product in a very traditional way of viewing commercial products. It is the artist manager's task to define the commercial viability of the artist as a product and to help develop and exploit all of the artist's talents in the marketplace. The marketplace in this instance is one where customers (fans) are able and willing to buy performance tickets, recordings, and merchandise.

Musical genre

Artists must be able to say in a few words what kind of music they perform and what audience they are identified with. They must be clear on the kind of audience they will appeal to and, if their music is to be pitched to radio, where the music is likely to be heard. The reason this is important for artists is that they will be asked continuously by people unfamiliar with them as an artist, "What kind of music do you do?" When artists have a keen sense of who they are artistically and commercially, they show a sense confidence to others that says, "I'm ready."

To the manager, it is important to have an understanding of the marketplace for the musical genre. Often, managers will seek artists to manage who are within musical genres they understand. But sometimes a manager will find a talented artist with commercial potential that isn't within the same genre. For example, Simon Renshaw managed the musical careers of the Dixie Chicks, Limp Bizkit, and Jennifer Lopez at the same time.

Biography

The biography should be the creation by a third-party professional such as a publicist, and it should be relatively inexpensive to produce. An experienced publicist will be able to put together the basics of a bio but will also be able to find the "back story," or discover things in an artist's background that distinguish him or her from everyone else who is trying to sing a song and become a star. At this point, the bio will already be a part of the press kit if it has been created. If it is not written, it needs to be done now because the background of the artist will be a key component of marketing communications the manager and others will use in the months ahead.

Talents

Including this section in the career plan will help the manager and the artist define the range of talents an artist currently has. This includes anything that the artist does as a performer or a songwriter that lets them demonstrate the range of their artistry. Make this section as detailed as possible, including any talents the artist may have developed in the past that are no longer used. For example, if an artist learned to play violin in grade school but hasn't played one in years, this is the place to note that for future reference because it might be redeveloped into a performance asset.

This section can also be the source of information that outlines talents the artist does not have. There may be some skills or talents the artist and manager feel should be developed, and including them in this section will help assure that they become part of the plan and find a place on the timeline for the future. For example, if the artist wants to learn to play the piano to help when he or she begins to learn songwriting, this is a good section in which to include it.

Experience

Assessing the experience of an artist can help both manager and client discover things in the artist's background that can smooth the transition into a quicker paced career. Many artists have had several years of seasoning preparing for a high-profile career, and some of those experiences can help plan how best to use that experience. Artists who began their own career in the music business have had to do most of the tasks associated with it for themselves, and that can be useful. For example, if the manager knows that the artist has done bookkeeping for a group he or she was in, it can make the creation of a career plan easier for the manager and business manager.

The range of experience for an artist can include:

- Performing as a front person or backup musician on stage
- Writing songs
- Playing instruments
- Managing tours
- Booking shows
- Creating musical arrangements
- Developing and planning stage performances
- Studio performance
- Bookkeeping
- Website creation and management
- New media management
- Press relations
- Producing recordings

Previous conversations with the artist might not have revealed the range of experience the artist has, and this part of the planning process will help to define the talents that are available to be exploited, and to begin planning how to use them to the advantage of the artist's career.

Uniqueness of the artist

Promotional planning in marketing departments looks for ways to spin a product as being "new and different," and then highlights benefits to consumers that will become part of the advertising message. It is not much different with the artist. In this section of the plan, the artist and manager will define the features of the client's artistry that make him or her unique, different, and relevant to commercial music. Often publicists can be helpful in this process because they are experienced in seeking the media value of potential stories in the music business. This section of the plan will also be helpful in refining the target market for the sale of tickets and recordings. There will be more about this later in the plan.

EVALUATION OF THE ARTIST

The previous section took an objective look at talents and skills that the artist brings into their career. This section provides a critique and evaluation of the artist. The intent here is not to criticize the artist, but to evaluate the artist with the purpose of finding ways to make the artist more competitive. Students of management will recognize this as a SWOT analysis, which creates an analysis of Strengths, Weaknesses, Opportunities, and Threats of the artist. Using this tool can give both the artist and manager a candid review of where the artist is as the plan begins, and it will also be helpful in defining goals later in this plan.

Strengths and weaknesses of the artist

This section of the evaluation of the artist looks at factors that are internal to the artist, meaning that they are often within the control of the artist and the manager to exploit (strength) and to improve (weakness). These can simply be lists of phrases that highlight both of these elements of evaluation. For example, strength can be "the artist has charisma in both their stage performance and in personal interviews," and a weakness can be "the group has had frequent personnel changes." Note that both of these are internal to the artist as defined previously.

Opportunities and threats

These are circumstances that affect or have the potential to affect the career of the artist that are external to the artist, meaning that they exist but are out of control of either the manager or the artist. In other words, the artist career plan must be structured so the artist will thrive because of the opportunities, and despite the threats. An example of an opportunity is "the musical genre of the artist is seeing a significant growth," and a threat might be "there are too many artists with a similar performing style." The challenge for the manager is to try to keep ahead of the threats that are brought about by the evolving business model of the industry and the continuous changes in technology, and the ways both impose on the earning streams of the artist.

Action points based on this evaluation

An action point is something that requires planning to turn it into an activity that will develop the career of the artist. When the SWOT analysis is complete, the manager and artist will have a list of points around which the development of the artist's career can be planned. The section of the plan that identifies goals and their strategies will rely on the implications of the SWOT analysis and will help the manager prioritize the timing of activities supporting the artist's career.

EVALUATION OF THE MANAGER

This part of the artist career plan is directed at the manager, and because of the nature of it, the manager may choose to omit it from the formal plan shared with the artist. In this section, the manager will give an honest assessment of his or her ability to manage an artist in this musical genre. It is a SWOT analysis for the benefit of the manager. As each new client is acquired and each career plan is created, the manager should generate this analysis as a continuing assessment of their growth as an artist manager. Artists in turn should conduct their own assessment of the manager using a SWOT analysis so they can have conversations with the manager about their planning based on what they have learned through the readings in this book.

Strengths and weaknesses of the manager

For the experienced manager of artists within a familiar genre, this section of the manager's evaluation will be relatively straightforward, and requires a measure of introspection. With the passage of time, the strengths will improve and the weaknesses should diminish. However, for the manager who does not keep current with trends in the roles of artist management, a candid review of weaknesses will reveal areas of necessary improvement. It will also reveal opportunities for the manager to consider taking short courses or attend brief seminars in management that will sharpen his or her skills.

In a competitive world like the music business, it is necessary for managers to continuously improve their knowledge and abilities to advance careers of their clients.

Opportunities and threats

This analysis will help the manager to develop a discipline of continually keeping abreast of the music business in areas that will be helpful to this specific artist. As mentioned frequently in this book, there is no area of the industry that remains static and it is imperative that the manager continuously reads all available information about the genre of the artists in the client list from sources like *Pollstar* magazine, *Billboard*, and available Internet sources such as *AllAccess.com* and *CMJ.com*. Listing the resulting opportunities and threats will provide an active and ongoing awareness of these matters. For example, if a manager learns that the latest recording contracts offered to new artists require that the artist must give up a portion of ticket sales from performances, the manager must develop a strategy with the artist's attorney to minimize the impact of this on the earnings of the artist. On the other hand, if the manager is working with a nonsongwriting artist who records for his or her own independent label and learns that the mechanical rate for licensing songs for recordings is increasing, it will be necessary to advise the artist that earnings from the sale of recordings will decrease if the price of those recordings remains the same.

Conflicts of interest

For every career plan, the manager should evaluate and disclose to the artist any potential conflicts of interest. There are the obvious conflicts noted earlier in this book, such as the manager owning a publishing company or a recording studio. As the plan is assembled, the manager should note any new conflicts of interest that have developed, such as managing or planning to manage artists who have similar talents as the artist for whom this plan is being created. The acknowledgment of any conflict of interest in this plan can minimize possible disagreements that might affect the relationship the manager develops with the artist.

ESTABLISHMENT OF GOALS AND TIMELINES

In this section of the career plan, the artist and manager discuss career priorities in the short term and the long term. The goals for the artist should be set in clear terms so that both understand them and both agree with them. The goals must be realistic and deliverable and the timelines for achievement must be realistic given everything else that is happening to support the artist's career.

Major goals for the artist and sample strategies and tactics to achieve them

A short-term goal might be for the artist to write songs to prepare for performances of original works. The short term in this case might be defined as within the next six months. As the plan is put together, a definition of terms such as "short term" is necessary to be sure the elements of the plan are realistic in the eyes of the artist.

An intermediate goal might be to perform some of these songs as an opening act for an established artist. Major steps to get the artist to this point will likely require recruiting musicians, rehearsing, booking shows to get experience performing, creating a recording to be available for fans, and perhaps securing a recording contract with an established label. A budget must be prepared to include funding sources to cover the expenses. It is not uncommon for an artist or a label to pay for the privilege of opening for an established artist.

A long-term goal could be to become a headliner. The artist and manager might see this as something that will happen in eighteen months or perhaps two years or longer. Steps to get the artist to this point can include local and regional touring establishing an increasing fan base, creation of a second recording with or without a record company, creating a bigger stage show, and seeking bookings at larger and larger venues. All of these points are ideas that have myriad details that become part of the strategies to support the goals themselves. Other longer-term goals could include things like getting a major recording contract, getting a publishing contract, getting a movie or television acting deal—anything the artist genuinely believes is possible and that the manager can support by creating a plan to achieve.

Goals supporting major goals (subordinate to major goals) and sample strategies and tactics to achieve them

Many of the lesser goals that the artist and manager include in the plan can be viewed as subordinate goals, meaning that when these goals are achieved, they will support a larger outcome. For example, an artist might want to learn to write songs, and it would become subordinate to a major goal of getting a songwriting publishing deal. Perhaps the artist wants to learn to play an instrument. This supports a goal of adding the playing of instruments into the artist's stage performance. A goal of getting studio experience could be a short-term goal that would support a larger goal of singing demos or recording an album.

Each of these goals requires a strategy with a timetable to assure that the artist stays focused and on target to achieve them. With the strategies formulated, there will be daily tasks called *tactics* that ensure regularly budgeted time and energy to support the strategies the artist and manager have agreed to.

Setting timelines

Putting the goals on paper and then pasting them into a visual timeline can be revealing. The goals must be identified by name, starting date, and completion date. When they are put into a timeline, it is apparent how time has been budgeted and how many things are happening at the same time. A timeline can reveal that there is too much or too little going on at certain times, and an adjustment is necessary.

As noted previously, the inclusion of a visual representation of the goals and their related tasks can be helpful in budgeting time, as well as eliminating the frustration of having too many or not enough things going on at the same time. A timeline for each goal should be created followed by a master timeline that incorporates the start and end dates for all major goals.

As timelines are prepared for goals, this is the time to attach a list of the gatekeepers who will influence the success of the goal. A gatekeeper is someone who has the ability to say "no" to your being able to advance through a portal along the career path. When setting up the goals and timelines, make note of who the gatekeepers are by title, by name, and by company, and include any contact information you can find about them. Also include the names of anyone within your network of professional contacts who might know the gatekeeper and can help you get past them.

The artist manager can find timeline software from companies such as Simplicity, which are relatively easy to learn and use. Older versions of their software are available at little or no cost on the Web and are very easy to apply to an artist's career plan. Internet sources for timeline software and other useful links are always available at this book's website, *http://www.artistmanagementonline.com*.

As you list gatekeepers, this is also the time to note checkpoints you will use as barometers to determine the progress with the goal. Checkpoints are nothing more than marks in your timeline that cause you to ask the question, "Is this where we thought we'd be at this time? If not, what should we do?" Equally important is the need to change any of the timing or strategies you are using to reach the goal. Remember, a plan is not something that is set in stone. It requires you to revisit it often to ensure the plan and its goals remain realistic. This means the artist and the manager should schedule at least a monthly review of the plan and its objectives to be sure the priorities remain the same, and then make any adjustments to it that are necessary.

Each goal will have its costs in terms of money or time. For example, if a goal is to write songs, it means the artist might not have time to do as much travel for paid performances. Indirect costs like this can be significant. Perhaps the goal is to learn to play piano as an added feature for the artist's stage performances. Learning piano has indirect costs, but it also includes payment to teachers and the cost of purchasing pianos for the artist to use for practice and on stage. As the manager puts together these costs, it is important that the artist understands them and agrees with the goal given its cost in both money and time.

DEVELOPMENT OF A MARKETING PLAN

Much of the work of an artist manager is being an advocate for the artists in the client list. In this circumstance, *advocate* is another word that means salesman and promoter. So it is not surprising that one of the key elements of the career plan includes brand management and the creation of a marketing plan for the artist. The plan requires a look at the business environment to determine the state of the industry the artist will become a part of and then looks at the target market, or fans, whose money will become the basis to support the artist's career.

The state of the industry

Using secondary research sources like trade magazines can provide the information a manager needs to evaluate the genre and generate ideas on how to position the artist to become a successful part of it. All things considered, it is important to know whether music and tickets sales are stable or declining.

It is important to know how recordings for a similar artist have been selling. For a new artist, for example, the manager will want to look at SoundScan data for new artists over the last year and see whether they have had sales success. Likewise, the manager will want to look at the *Pollstar* data for ticket sales for performances for the same reasons. Because every artist is different, and every artist is not presented in business the same way, looking at historical sales data can be used only in a scenario of "all things being equal." Perhaps other artists were poorly managed, or may have had recordings that did not appeal to buyers, or perhaps could not get bookings necessary to support themselves. Look at available data, but use them only as reference points.

The review of an artist's business performances within the genre also reveals the number of other competing artists. As it discloses the competition, it also gives the manager information to find opportunities for his or her artist.

Any review of the state of the industry would be incomplete without a look at how radio is using music within the genre. Although an artist may have a manager but not a major record deal, it is still important to know which markets for tour stops have support for the music on terrestrial radio (radio with broadcast towers). If radio does not support the music, promotion of a tour will become very different from one in that radio is a supporting element. Arbitron, radio's audience measurement service in the United States, gives regular updates on radio stations by market and by genre at its website, *http://www.arbitron.com*.

The manager should also look at opportunities in international markets. Certainly the first markets that English-speaking artists should look toward are the United Kingdom, Australia, Canada, and the United States, all of which are known to support the purchase of recorded music and concert tickets. However, there are instances where acts like the jazz vocal group Take 6 have continuing commercial success in Japan and Brazil. It is up to the artist manager to seek opportunities like these for the artist, determine whether it is a practical use of their time, and then present the ideas to the artist for consideration.

The target market for the artist

Earlier in the book we discussed target markets, their attributes, and why they are important to the manager and the artist. In this section of the artist career plan, it is important to specifically define the target group that both the manager and the artist feel will be likely purchasers of the artist's recorded music and tickets to his or her performances. Here is where the artist and the manager create the narrative description of all possible consumers.

To make a single description would make the target too narrowly focused. If this plan exploits all available talents of the artist, then the artist will have a continuously broadening appeal, and the result will be a growing fan base and a broadened definition of consumers and fans of the music that were not part of the original target. Remember that a career that develops a four-year history may now have a fan base that is no longer in high school, but may be becoming seniors in college; or, a 27-year-old junior executive might now be a 31-year-old who travels frequently. Suddenly it is no longer basic demographic information that defines the consumer; rather, it is now lifestyles and values of the consumers, and the manager should plan for these changes and to be aware that they will happen. The reason many artists' careers are declining after two or three years is because fans and pop culture have moved on to new music and left those artists behind. This part of the plan acknowledges the likelihood that this will happen and includes the commitment to stay current and relevant.

Likewise, four years into a career will include the creation of new ways for the consumer to use the artist's music. Technology advances at a rapid pace and the migration of music into new platforms and media is much quicker now than in the days of vinyl, 8-tracks, cassettes, and CDs. The plan in this section must acknowledge the likelihood that these changes will happen, and both the artist and manager commit to keeping the target consumer clearly in mind by considering the social, demographic, uses of kinds of media, and other changes that will happen to the fan base.

So the structure of this section of the plan includes a very specific description of current and anticipated future fans of the artist. This is not only helpful to the artist and manager in their planning, but also to concert promoters who will employ the services of the artist.

Detailed plans to reach the target

Creating this section of the plan will require combined input from the team the artist and manager assemble on behalf of the artist's career. The team will consider the goals set out by the artist and will then begin constructing all of the details necessary to create a reasonable plan to get the artist in front of those willing to buy performance tickets, merchandise, digital sound accessories, and recordings. The next section of the career plan will briefly look at the responsibilities of those who make up the team and it outline how and why their participation in this section of the career plan is important.

This part of the plan will necessarily be the largest part of the document. It will be replete with the daily tactics necessary to support the strategies to achieve the goals. For the manager, this information will continually be a reminder of where the management of the career should be from day to day and week to week. For artists, it is a continuing guide to let them know the responsibilities shared with the manager to achieve success.

This section of the plan will likely be the part that requires the most revision. Implementation of the career plan is done within the reality of the business environment, and within the continuing changes of the music business requiring that the plan be modified monthly, perhaps weekly, to reflect that reality.

BUSINESS FRAMEWORK
Form of the business

In consultation with the business manager, the accountant, and the attorney, a recommendation will suggest the most appropriate business form for the artist. This will require a forward look at the goals of this plan and their timing. A simple business form may be adequate for the short term, but as the artist begins to have more elaborate business relationships, it will be necessary to set up the business in a form that may resemble a corporation, a formal partnership, or a limited partnership or company. In the United Kingdom, the decision is whether to be a sole trader, a partnership, or a limited liability company, but the attributes of these forms of business entities are very similar to those in the United States (HSBC, 2006).

Personnel requirements

At the beginning of the artist's career plan, there is a need for initial advice from members of the team in order to launch the business-related activities of the artist. The advice is for planning and budgeting purposes, with their individual areas of expertise becoming a regular part of the support group for the artist. For example, a publicist may initially be consulted to update and refine some of the promotional materials needed for the artist, but they may later become a continuing part of the artist's team to support touring and other matters.

The professionals who may need to become part of the team could include a publicist, a manager of new media, an attorney, a banker, an accountant, a webmaster, an insurance agent, a business manager, a tour manager, an agent, an image consultant, a show designer or producer, a wardrobe and makeup consultant, a band or backup group, and others. As each of these is introduced into the career plan, a note should be made identifying the costs, if any, that are associated with their services to the artist as well as the timing of when they will have an impact on the budget.

Insurance

The plan should include the necessary insurance based on the experience of the manager and advice of the insurance agent serving on the team. Some of the insurance coverage discussed earlier in this book will not initially be necessary, but it should be put into the plan with an anticipated date on which it will be required, along with an estimate of how much it will cost. For example, if the artist will have a developmental period that will not include touring, broad liability insurance coverage will be unnecessary. However, the manager must plan for the time when it becomes necessary because it creates a cost when touring begins.

Other

Within the business framework of the plan, the manager may suggest that it should include items such as memberships in unions and other professional organizations, subscriptions to publications in print and on the Web, and associating with a performing rights organization.

THE FINANCIAL PLAN
A personal budget for the artist

The example created in Chapter 6 is a good template to use for this section of the artist's career plan. Without this element as part of the plan, the work of the manager could become compounded as it becomes necessary to also manage the artist's personal finances in addition to all of their other responsibilities. This budget should be an annual budget broken down by month, and it is important that it be a reflection of the standard of living to which the artist is accustomed.

Budget for career plan

Budgeting is not about determining how much money the artist can earn in the next year; budgeting of any kind is actually the process of setting priorities. Within the limits of expected income, the artist and the manager must decide what the most important use of available money is in order to meet the career objectives of the artist. In other words, how do we use these earned resources to meet the goals of this plan?

The budget the manager prepares in order to support the artist's career plan should be for a three-year time period. The first year will have considerable certainty, whereas the third year will have very loose estimates of earnings and expenses. All of the sections of this written plan provide important information about the timing and amount of income, and the timing and amount of expense. The budget for the first year should show income and expenses by the month so that the manager can track the available resources to pay for the plan and perhaps postpone some expenses when income doesn't match the timing of the income. When the monthly budget estimates are complete, they then become the input for the annual budget with the 12 monthly totals combined. The second and third years should be broken down by quarters, again showing the timing of the income and expenses.

It is important to acknowledge revenue and expense timing in the budget. Unless there is a considerable monetary cushion for the manager to rely on, spending earnings before they are earned can create an annoying distraction because the manager must scramble to find ways to cover expenses rather than stay focused on the artist's career. Some of those major expenses can include the personal budget for the artist, travel expenses, personnel costs, recording costs, plus all of the costs associated with this career plan.

EXIT STRATEGY

An *exit strategy* is the plan a manager will use when the business relationship with the artist needs to come to an end because the business simply isn't there anymore. It doesn't necessarily mean an end to a personal relationship, but happens when the manager finds that his or her time dedicated to the business of the artist is no longer profitable.

An exit strategy doesn't have a timeline. There is no specific date on which an artist manager will implement an exit strategy. Circumstances put an exit strategy into action. For example, it becomes clear that an artist has lost mass market appeal and it is time to direct the artist's career into areas where all can profit from past successes. Likewise, the manager should consider the certainty that the relationship with the artist will change, and consider the possible circumstances that will require the association to change.

The artist in a mature career

The artist career plan must assume the differences from a career in its prime and a mature career. Careers in their prime will include elements of building fan bases, regular touring (perhaps in larger and larger venues), and issuing new recordings on a regular basis. If the artist and the manager plan for the reality that a career nearly always goes into decline, developing a vision for profiting during the mature career will help prepare for it when it arrives. Mature careers for many artists are a mere three or four years beyond their prime. For others, it may be several years longer. In either case, the artist and the manager should be collecting all of the assets of the career in its prime for use later. For example, the manager should build an audio and video archive of performances and recordings, beginning with the earliest demos. These items should be protected from deterioration and kept somewhere safe. They could be invaluable as the artist relies on these periods of success to fuel later stages of the career.

Too often, artists who achieve a measure of success begin to think it will last forever, but it rarely does. It is a shock when fan attendance at performances declines, and it is surprising to the recording artist who finds that their current recordings aren't selling on iTunes. Both are signals that things are changing or are about to change. Use this portion of the plan to take a long-term "what if" look at the artist's career, and consider what can be done now to prepare for it.

Planning by the manager to end the relationship

We have referred to the nature of the artist–manager relationship as being like a marriage. It is at the end of the business relationship where it genuinely becomes different. When it is clear that the future of the business between the artist and the manager is ending, the manager must objectively view it for what it is and then break the tie. Time is the most valuable resource an artist manager has, and if that time invested on behalf of the artist is being unproductive with little chance of that changing, it is time to suggest new management to the artist. It does not mean that the manager breaks a friendship; rather, it just means the nature of the relationship changes.

Presidents of record labels are often fired by corporate ownership, replaced by someone new, and the new president quickly pares the label roster and fires unneeded employees. The approach is easy and clean this way because the new president has no relationship with artists whose recordings no longer make money for the company, nor does he or she have a relationship with surplus employees. That is how the corporate world does it. The artist manager, on the other hand, must personally end the relationship with artists who are no longer productive to himself or herself as their manager. It is much more difficult because it is much more personal. Making the decision and acting on it is important, and it is never pleasant. If a manager is slow to act, the nonproductive artist drains time and energy from the manager and the company.

For the manager, consider what criteria will be used to decide when it is time to move on to focus on other clients. Perhaps the manager will want to consider when it is time to cut losses incurred by trying to keep a declining career alive. And the most important "what if" that a manager should always be aware of is the possibility that the artist will choose to move on to another manager. Never take the contract to manage an artist for granted. If a manager does, there will always be the chance that they will be blindsided by the artist who wants quicker career acceleration and will move on to a new manager. An entertainment attorney specializing in artist–manager contracts once said that recording artists who achieve measures of success sometimes replace managers after three or four years—not because they are ineffective, but because they remind them of the days when they were struggling to make a place for themselves in the music business. The manager should plan for an artist to change management at some time, which can help to minimize surprises to their own career, and to always have an exit strategy available.

This exit strategy section of the artist career plan is one that the manager should keep as part of his or her private copy of the plan. Failing to include its consideration may mean that the manager will need to react to a sudden change rather than being ready to deal with possibilities he or she should have already considered.

THE PLAN OUTLINE

Here is an outline of the artist's career plan that can serve as a template to create a working plan for an artist. The plan includes some prompts to assist the manager in assuring relevant considerations are included in the plan. As this is a template, it is only a beginning point; other personalized plan attributes must be added that are unique to the artist and their career goals.

Recording artist business plan

I. About the artist
   A. Musical genre
   B. Biography
   C. Talents
      1. Sing
      2. Play
      3. Write
      4. Perform
      5. Other
   D. Experience
      1. Writing
      2. Singing
      3. Performing
      4. Other
   E. How the artist is unique from others
      1. What is it about the artist that should be promoted to set him or her apart from others?
II. Evaluation of the artist
   A. Strengths, weaknesses of the artist
   B. Opportunities and threats external to the artist
III. Evaluation of the manager's ability to manage an artist in this musical genre
   A. Strengths, weaknesses of the manager
   B. Opportunities and threats external to the manager
   C. Real or perceived conflicts of interest
IV. Establishment of goals/comprehensive timeline for this plan
   A. Major goals for the artist and sample strategies and tactics to achieve them
      1. A record deal
      2. A publishing deal
      3. A tour
      4. A movie deal
   B. Goals supporting major goals (subordinate to major goals) and sample strategies and tactics to achieve them
      1. Learn songwriting
      2. Learn to play an instrument
      3. Develop a stage show
      4. Perform in smaller, then larger venues
      5. Get studio experience
   C. Timeline for each goal
      1. Beginning and projected achievement dates
      2. Gatekeepers (all by position, name, and company)
      3. Checkpoints or significant steps between beginning and end of timeline for each goal
      4. All direct costs that can be anticipated that are unique to each goal

V. Development of a marketing plan
  A. State of the industry
    1. A general evaluation of the genre
    2. Sales of recordings within the genre
    3. Sales of tickets within the genre
    4. Number of competing artists within the genre
      • Opportunities for new artists within genre
    5. Competing artists
    6. How the radio format is doing in audience ratings
    7. International opportunities there for the genre
  B. The target for the artist (those willing to buy tickets and recordings)
    1. Demographics
    2. Lifestyle
    3. Generational attributes
    4. Uses of entertainment
    5. Uses of music (how it is consumed)
    6. Values of the target
  C. Plans to reach the target consistent with goals
VI. Business framework
  A. Form of the business
    1. Proprietorship, partnership, corporation, subchapter S corporation, LLC, LLP
    2. Personnel requirements, when they should be employed (by calendar quarter), and an estimated cost:
      • Publicist
      • Attorney
      • Banker
      • Accountant
      • New media manager and webmaster
      • Insurance advisor
      • Business manager
      • Tour manager
      • Agent
      • Image consultant
      • Show designer/producer
      • Wardrobe consultant
      • Band
    3. Costs associated with personnel, such as the investment in a website
  B. Insurance
    1. Life, vehicle, liability, equipment, health
  C. Other
    1. Performance Rights Organization (PRO), union memberships, National Academy of Recording Arts and Sciences (NRAS)

VII. The financial plan
  A. Personal budget for artist
   1. A complete annual budget for the artist that anticipates living expenses and sources of income for the coming 12 months
   2. Budget should reflect the standard of living to which the artist is accustomed
  B. Budget estimate to support this business plan
   1. Budget is for three years:
       • First year is monthly
       • Second and third years are by quarters (90-day periods)
       • Each annual budget expresses the financial goal for the year
   2. Budget for each period shows revenue to include:
       • Sources of income
       • The timing of the anticipated receipt of the income
   3. Budget for each period shows expenses to include things such as:
       • Personal budget for the artist
       • Personnel
       • Travel
       • Merchandise
       • Recording
       • All other plan elements that have costs associated with them
VIII. Exit Strategy
  A. Plan by manager to wind down and end the relationship
   1. At what point do you cut losses?
   2. At what point do you move on to more lucrative projects?
   3. What do you do if the artist ends the relationship?
  B. Plan for the artist to thrive in a mature career
   1. Collect resources during career for use in maturity
   2. Design strategy for smooth transition

Reference

HSBC. (2006). *HSBC UK Business*. http://www.hsbc.co.uk.

Coaching, leadership, and final advice

13

When you think of coaches, images often come to mind of individuals who are icons of strategy and execution in the sporting world. They understand the game they're playing, they know the level of the talent available to them, they have a plan to compete, and they've cleared another place on their trophy shelf. Their specialized experience and understanding of human nature help them build on an individual's talents and prepare the individual to compete. The role of the manager as a coach, especially early in an artist's career, is very much like that of any other coach.

Likewise, we look at leaders as people who have that special gift of being able to motivate people to become involved in ways that benefit their collective and common good. They are agents of change who can express a vision that draws people to them and inspires them. Crises are times when leaders step up and involve others to act for a solution. When things are falling apart, it becomes a time when we look to a leader to become involved. Again, leadership in a manager is important to new artists who need help directing their budding careers.

The following discussions about coaching and leadership are intended to present insight into those who successfully coach and lead. They examine some of the qualities of those who find themselves as coaches and leaders so that the artist manager can develop a strategy to build those qualities in themselves. Knowing

161

what coaching and leadership are about gives a basis to observe those qualities in others, and then allows the manager to adopt those features that complement his or her style when working with artists.

COACHING

A coach has been defined as a business mentor, a motivator, a trainer, a tutor, and an accountability partner, but for our purposes we'll define an artist manager/coach as someone who is seeking coaching outcomes that will improve the artistry of the artist, and help him or her reach career goals.

Depending on the managers' backgrounds, their areas of expertise will define the kinds of coaching that they can provide. Sometimes managers will be the coach; sometimes they will seek coaches to help their artists. The earliest coaching a manager will give artists is to help them with the business etiquette of the music business. There will be innumerable meetings, gatherings, luncheons, dinners, conventions, showcases, and other functions at which artists must be coached and guided by the manager on how to handle themselves. There will also be countless public performances that will require critiques followed by feedback and coaching for improvement. The manager will also be asked by the artist for coaching in those specialized areas in which the artist manager has background, knowledge, or expertise.

Author James Flaherty created one of the best books about coaching, and ideas he presents as adapted for use by the artist manager can provide some insight to the work of the manager–coach.

First, coaching is a very personal thing and there is no reference one can use to provide coaching in a specific circumstance. People are all very different with different backgrounds and values, and they are in different stages of their lives. So there isn't a series of bullet points to offer a coach to be effective in a specific circumstance. Despite the previous observation about bullet points, there are three products of coaching:

- The artist manager–coach seeks long-term quality in artistry.
- The manager seeks the situation in which the artist learns to recognize when he or she needs to correct something, and then do it without prompting.
- The manager seeks the situation in which the artist takes the initiative to find ways to grow artistically.

These three products or objectives of coaching take the manager out of the role of coach until another opportunity presents itself.

Coaching opportunities for the artist manager can occur: (1) when artists request or need a review of aspects of their creative performance, (2) when they have a disastrous setback in their career, (3) when they ask the manager for assistance as their coach, and (4) when they need a new skill such as playing an instrument or learning to write songs. Areas where Flaherty's book suggests that coaching would be difficult, if not impossible, include issues dealing with social identity and habits the artist has acquired. Habits that require considerably more than coaching are those associated with substances. Artist manager Joyce Moore tells of her challenges of dealing with the drug habits of singer Sam Moore, saying she "got control of his money, his drugs, his proximity—I had to play hardball, tough love" (Moore, 1993). And her approach was successful.

The foundation of coaching between an artist and manager is the relationship the two share. It has to be built on mutual trust. The artist must genuinely believe that the manager will be discreet where necessary and candid when required. Likewise, the manager must believe that the artist is open to being coached.

The relationship must also include mutual respect with a mutual freedom to express. The artist doesn't need to respect the manager on all matters; rather, the manager must coach in an area in which the artist views the manager as someone qualified to help the artist acquire the desired competency. Freedom to express within the coaching relationship doesn't mean each can say anything to the other without consequences; rather, it means both agree to genuinely listen to each other in the coaching environment and agree to express in total confidentiality.

Coaching conversations come in three forms. The first is one in which only a single conversation is necessary. This is where the artist requests coaching and the manager provides it. This conversation is also about standards of performance for things like writing and performing and can begin with the question, "How do you think I'm doing . . . ?" Or the first kind of conversation can begin with the manager pointing out mistakes the artist is making and offering ideas of how they could be corrected. These conversations are advisory as well as coaching in nature, and Flaherty labels these Type I coaching conversations.

The second kind of conversation, a Type II conversation, requires more than one meeting on the issue. It is one that begins with an opportunity to coach, but it calls for at least one follow-up conversation. For example, if an artist is disorganized on stage, doesn't have a performance set list for the band, hasn't memorized words to new songs, and doesn't know where he or she should be positioned during performances, there is an opportunity for the manager to coach the artist on strategies to prepare for performances. Subsequent conversations will be needed to assure that the artist has accepted the coaching and is indeed organized for performances.

There is a third kind of coaching conversation offered by Flaherty, but it deals with fundamental change, an area that would typically not subject to artist manager coaching.

When the artist manager has established a coaching relationship, there are ways to strengthen it. One way is to tell the artist you will do something, and then do it. Following through on a commitment in the name of coaching the artist is a solid strategy to underscore how serious the manager is in the success of the relationship. Another way is to bring the artist into decision-making processes that affect the area being coached. When artists see the amount of care and consideration necessary to make decisions—and become participants for a period of time—they will clearly see what it takes to direct their careers. And finally, a strategy that will create a stronger coaching relationship is one where the manager tells the artist about how his or her opinion was changed by someone else through coaching or mentoring.

Special thanks are extended to author James Flaherty and Elsevier/Butterworth-Heinemann for permitting the adaptation of ideas from the book entitled *Coaching: Evoking Excellence in Others* (2005) to this subject of artist management.

LEADERSHIP

Leadership is one of those qualities a person has that genuinely makes him or her stand apart from others. Leaders are the kind of people who quickly step up to a challenge or opportunity and say, "I'll take care of it," and then they do. Leaders present themselves in boardrooms, in the military, on football and soccer fields, in civic clubs—anywhere someone is needed to coordinate and direct the energy of a group or an individual to meet goals. What follows is a brief background of leadership, including some working definitions that may be useful for one who wants to understand and develop the skills and traits of leading.

The typical view of a leader is someone who is able to influence followers to act, even if "followers" is limited to one person. An artist manager has a special position of influence on the new artist who is launching a career with the manager's guidance, and this often defines the leadership role of a manager as being the career coach for the artist. The manager also takes on the role of leadership and influence with the team of professionals and specialists that supports the artist's career.

Leadership skills can be developed by taking on increasingly important challenges and successfully completing them. In the common team environments that many offices and businesses use today, there is opportunity for a budding leader to step up and agree to be responsible for the completion of a project. For someone new at taking on the responsibilities of leadership, it will seem risky at first. Certainly there is risk of failure, but with good planning and adequate support, these early opportunities can give invaluable experience in developing the qualities of leadership. Much of the most effective leadership training involves experiential simulations, and the best leadership training happens when the individual can become involved with a real project in a leadership role. Educators often cite the value of theory coupled with experience as one of the best ways to inspire genuine learning. Earlier in this book, we discussed how one can begin work in the profession of artist management, and a suggestion was to join an existing firm as a junior associate—perhaps as an intern or apprentice—and be placed into circumstances where elementary leadership skills can be developed.

The term "leader" is often defined by the names of those who lead because we have a sense of the traits or characteristics the individuals possess that make them effective in that role. An ongoing survey by Kouzes, Posner, and Peters lists characteristics respondents would favor in those they would choose as leaders. In descending order, the characteristics are:[1]

Honest	Courageous
Forward-looking	Cooperative
Inspiring	Imaginative
Competent	Caring
Fair-minded	Determined
Supportive	Mature
Broad-minded	Ambitious
Intelligent	Loyal
Straightforward	Self-controlled
Dependable	Independent

The group has found that the top characteristics people seek in their leaders remain relatively consistent from survey to survey.

Leaders also have been found to have personality traits that make them effective leaders. Hughes, Ginnett, and Curphy assembled a list of those "consistent patterns of behavior" for their book *Leadership: Enhancing the Lessons of Experience*:

- *Dominance*—Defining this trait are phrases like "confident, forceful, outspoken, and opinionated."
- *Self-confidence*—These are persons who "feel comfortable with their own judgment, abilities, or skills."
- *Achievement orientation*—Leaders who are highly achievement-oriented "complete tasks and activities primarily for the satisfaction of accomplishing a challenging goal." The authors say they "tend to be hardworking, ambitious, and competitive."
- *Dependability*—Leaders high in this quality "tend to be conservative, careful, responsible, and reliable."
- *Energy and activity level*—Leaders who have a higher energy and activity level tend to be the most successful leaders.
- *Self-monitoring*—This quality of leadership defines how much a leader is willing to modify his or her leadership style dependent upon circumstances.
- *Locus of control*—This is a reference to the leadership style used to control followers. An external locus of control uses "coercive power," whereas an internal locus of control uses rewards and the expertise of the leader.
- *Tolerance for ambiguity*—This leadership quality suggests that leaders by nature are comfortable with "unstructured problems or uncertainty."

---

[1]Kouzes, Posner, and Peters (1996).

- *Adjustment*—This quality says that leaders tend to be well-adjusted and able to handle stress.
- *Sociability*—The authors say leaders with this quality are "outgoing and socially adept and tend to exert greater influence in a group."
- *Agreeableness*—Research shows that leaders with this trait tend to be more effective than those who are moody or intolerant.[2]

Another quality of effective leadership is that of being a visionary. A *visionary* is someone who sees things as they should be, develops a plan to make the vision a reality, assembles the resources necessary to succeed, and draws followers to share the vision who will see it to its achievement. Although we know the artist career plan is a joint vision of the artist and the manager, it is the manager who pushes the vision toward the biggest and most realistically attainable goals the artist can achieve. As a leader in this environment, it is necessary to manage the expectations of artists, which means they must understand that their grand and shared vision will have periods of time during which their career activity becomes slower than at other times.

Jack and Suzy Welch add a few other ideas that work well in the business world of the artist manager. They say that leaders should mentor those who follow by giving feedback, suggesting ways to improve, and congratulating successes. They say that candor is "one of the defining characteristics of effective leaders" (Welch, 2006, 120). From this author's perspective, complete but tactfully measured candor by the manager/leader with an artist is essential.

FINAL ADVICE

The final two points in this book are found in bits and pieces throughout the text, but their importance to an artist's success requires restatement.

First, when dealing with a record company that has the resources to be able to pay an artist an advance, seek the largest amount you can get from the company. An advance is merely prepaid royalties for the artist's recorded music project, so it will be ultimately charged back to the artist for recoupment. There are several reasons the manager should seek a large advance:

- Advances can help offset the startup costs of preparing artists for their career and cover living expenses.
- They can serve as an interim income stream between the time the album is recorded and released and the time that royalties from sales actually begin to flow to the artist. This gap can be two years or longer—maybe never—and banking a large advance early on can cover career development costs.
- A large advance represents a significant commitment by the label to the success of the artist's recording, because if the project isn't successful, the label has no

---

[2]Hughes, Ginnett, and Curphy (1993).

chance of recovering the money they advanced to the artist. We discussed earlier how all artists at record labels are not treated equally, so a large advance can help assure equal treatment for a manager's artist. Likewise, a large advance from a publishing company demonstrates the commitment by them to the artist's songwriting.

- A large advance to an artist will help a manager begin to recoup the expenses they have had to absorb associated with launching the artist's career. Advances of this nature to an artist are immediately commissionable to the manager based on the terms of the artist–manager contract.

The second piece of advice seems obvious but certainly deserves restatement. Find the absolute best music you can for your artists: signature songs. Nothing satisfies a music consumer like a completed album to download that is filled with masterful songs and performances, and finding a signature song for the first album can quickly accelerate a career and build a lifetime of earnings for an artist. A record company's A&R department and the artist's producer will find music for the recorded music project, but as a manager you can't assume they will find the best music available that fits your artist, especially in this era of smaller label budgets and staffing. Be involved in the album production planning by refusing to accept the "acceptable." Be patient, and find the best music. Certainly the record company has every incentive to create a good album, but the author recalls a business seminar in which veteran producer James Stroud said that one of his artist-songwriters puts only two hit songs on a ten-cut album, and saves other potential hit songs for the next album release. It is unclear how that strategy has been working for the artist because he did not identify who it is. Successful enterprises always exceed the expectations of their customers, which is a sound strategy for the music that artists memorialize within their recordings.

References

Flaherty, J., (2005). *Coaching: Evoking Excellence In Others.* Butterworth-Heinemann/Focal Press, Burlington, MA.

Hughes, R. L., Ginnett, R. C., & Curphy, G. J. (1993). *Leadership: Enhancing the Lessons of Experience.* Burr Ridge, IL: Richard D. Irwin, Inc. pp. 150–156.

Hughes, R. L., Ginnett, R. C., & Curphy, G. J. (2002). *Leadership: Enhancing the Lessons ⌐ Experience.* New York: McGraw-Hill Higher Education.

Kouzes, J. M., Posner, B. Z., & Peters, T. (1996). *The Leadership Challenge: ¬ Keep Getting Extraordinary Things Done in Organizations.* Hoboken: Jossev Wiley & Sons.

Moore, J. (1993). *Soundtrack of Only the Strong Survive: A Celebration ⌐* CA: Miramax Home Entertainment.

Welch, J., & Welch, S. (2006, January 30). The Leadership Mindset

Take 6 marketing brief

Professor and artist manager Chris Palmer provided this copy of 60 days of activity he directed around the release of the Take 6 album *Feels Good*. It is presented here as an example of the immense energy and coordination required by an artist manager to successfully launch an album on an independent label. The author extends special thanks to Mr. Palmer for sharing his work for the benefit of those who are learning about artist management. Professor Palmer teaches at the University of Miami in its Department of Music Business and Industry.

TAKE 6 MARKETING/PUBLICITY REPORT 4/1

Gospel Music Week – Participating in Central South showcase and two full days of interviews.

Syndicated radio (Urban & Gospel)

XM's *Music Mondays* wants Take 6 on show. Date to be determined.

XM Radio – Watercolors – adding in rotation – various tracks.

XM Radio – The Spirit – adding in rotation – various tracks.

BeBe Winans Radio Show – interview completed on 3/24. Spots running now, interview set to run with the next week or two.

Gospel Entertainment News – interview completed on 3/24, running 2nd week in April.

Radioscope – interview completed on 3/24, airs the weekend of 4/15.

Bloomberg News – interview on KBLX in San Francisco on 4/3. Goes to 40 stations in May & June for Black Music Month.

Syndicated radio recently aired

XM Satellite Radio – live interview aired on 3/23 for "Greatness by Design"

ABC Radio (Rejoice Musical Soulfood) – live interview on 3/23.

The Tavis Smiley Radio Show (UPI) – interview completed on on 3/24.

SGN The Light – interview completed and aired on Broadcasting.

American Urban Network Radio – interview completed

169

Radio

"Do You Know What It Means" digitally serviced to Gospel, Jazz, Urban Adult Contemporary, and Mainstream Adult Contemporary radio formats.

"More Than Ever" and "Come On" serviced to Urban AC and Gospel.

WPZE Praise (Atlanta) – will schedule when the guys are in Atlanta.

V-103 (Atlanta) – working to get the guys on 4/24.

KHVN (Dallas) – booked for 4/25.

KSoul (Dallas) – booked for 4/24.

WMUZ (Detroit) – Cedric is scheduled to be on the air live 4/28. Morning drive.

WVOF (Fairfield, CT) – *www.upperroomwithjoekelley.com* – playing "Family of Love."

WFDU (New Jersey) – interview completed for their Gospel Grooves show, airing on 4/11.

Radio recently aired

KJLH (Los Angeles) – live interview with Cliff Winston aired on 3/21.

WJLB (Detroit) – live interview completed on 3/26.

WHCR (Voice of Harlem) – attended NY show.

WRSU (New Jersey, the Power Gospel Hour) – interview completed on 3/28, radio spot aired following morning.

WYCA-Gospel (Chicago) – live on the air with the whole group for one hour on 3/21.

WYCA-Urban (Chicago) – on the air live 3/21.

KISS- FM (New York) – live interview completed on 3/26.

Launch Radio Network – interview completed on 3/22, spots ran 3/22 & 3/23.

WKNC (Raleigh-Durham, NC) – a cappella program broadcast 3/19 – playing various tracks from CD on weekly Sunday morning a cappella program.

Print

Gospel Today – CD review May/June issue.

Xii Magazine – CD review running in their Spring/Summer issue.

L.A. Focu – mention in March issue, CD review coming in April.

Heart & Soul – CD review running in April/May issue.

Urban Network/Urban Inspired – interview completed on 3/24, feature running in April issue.

Precious Times – running CD review ... not sure if it was able to make the ˙ng issue.

˙ired Living Magazine – interview completed on 3/24 and will run in their ˙sue.

˙tts Times – CD and Concert review running in April.

The Wave Newspaper – interview and feature completed, waiting for run date.
The LA Jazz Scene – concert and CD review in May issue.
In Fuze Magazine – *www.infuzemag.com* – Article confirmed for April issue.
Performing Songwriter Magazine – review forthcoming could be 2–4 months.

Print recently run

Billboard – CD review ran in 3/18 issue.
   Times West Virginian – piece ran 3/16.
   Radio & Records – release date piece 3/21 issue.
   Indianapolis Star – *www.indystar.com/apps/pbcs.dll/article?AID=/20060226/
ENTERTAINMENT04/602260339/1081.*

Television

BET's "Lift Every Voice" – on board to do a profile with Take 6. Tape 4/14 air 5/14.
   Gospel Music Channel – using pieces of "Feels Good" in background commercial plugs.
   CNN – booked to be on live Sunday, April 23rd at 9:30 a.m. (EDT).
   Good Day Atlanta – scheduled to tape on 4/24 in Atlanta.
   At Home Live (Ft. Worth, TX) – would like to have the guys on 4/25 or 4/26.
   The 700 Club – they've done this show in the past, checking taping dates to see what can work.
   CBS Early Show – group will be live on 4/15.

TV recently aired

TBN Chicago – live interview on "Behind the Scenes" ran on 3/22.
   ABC Morning News Chicago – live interview and performance 3/21.
   Smooth Jazz – Jazz program ran on airline TV programming as well as in the following markets: Austin, Baltimore, Boston, Dallas, Denver, Detroit, Fresno, Harrisburg, Hartford, Houston, Indianapolis, Long Island, LA, New Orleans, New York, Philadelphia, Providence, San Antonio, San Diego, San Francisco, Seattle, Springfield. Also included radio advertising for the show in these markets.

Internet

Urban & Gospel

Gospelflava.com – CD review ran 3/28 and feature set to run within next two weeks.
   Gospelcity.com – feature set to run 4/5 or 4/6. E-blast from L.A. dates on site now.
   EUR/Gospel EUR – feature runs the week of 4/2.

The Belle Report – various e-blasts have run on 2/16, 3/21, and 3/27.

ChristianHangSuite.com – interview completed on 3/24.

Urban Network.com – running concert review.

The *ICN.COM* – interview set to run in early April.

Detroitgospel.com/Chicagogospel.com – e-blast on CD ran 3/21, contesting running through 4/21.

ChristianMusicToday.com – CD review and feature ran on 3/27.

Artist Launch – *www.artistlaunch.com* – "Come On" – featured on front R&B page March 9.

Flash Magazine (in Germany) – *www.flashmag.de* – review coming in April issue.

Soultracks – Promotion to preorder CD went out in newsletter and on front page of site 3/14.

Swissgroove – *www.swissgroove.ch* – featuring CD on their "SmoothVibes" Channel plus review.

24 Hour Gospel Channel – *www.24hourgospelnetwork.org* – posting review on 3/21.

Artist Launch – *www.artistlaunch.com* – Lamb Of God – featured on Christian/Spiritual chart.

Gospel City – *http://www.gospelcity.com/dynamic/music-articles/new_music/262.*

Jazz

About Jazz – *www.jazz.about.com* – profile coming by April 1.

Contemporary Jazz – *http://www.devcontemporaryjazz.com/thebuzz/?p=127.*

All About Jazz – *http://www.allaboutjazz.com/php/news.php?id=9204.*

E Jazz News – *http://www.ejazznews.com/modules.php?op=modload &name=News&file=article&sid=5836&mode=thread&order=0&thold=0.*

Jazz Monthly – article to be in their May release *www.jazzmonthly.com.*

Jazz Review – *http://www.jazzreview.com/cd/review-17620.html* – review from Karl Stober.

Jazz Review – profile feature on website for April 1.

Acapella

Acadisc – New info added on band at *http://www.acadisc.com.*

All Things Acappella – Interview conducted with Cedric on March 12 for special radio piece.

Contemporary Acappella Society Association – *www.casa.org* feature.

A-cappella – *www.a-cappella.com* Aaron Neville press release to their 22,000 email list 3/14.

E-blast announcing "Feels Good" at *www.a-cappella.com* 3/21.

Internet radio

Totally Acappella – playing various tracks – *http://www.totallyacappella.com.*

Acappella Radio – *http://www.casa.org/index.php?option=com_content&task =view&id=19&Itemid=47.*

Airplay Direct – mp3 download service exclusively for radio programmers/ music directors only. Yahoo! Music – added various tracks.

AOL Radio – added tracks into Gospel and Smooth Jazz programming as of 3/1/06.

iTunes Radio – added "Do You Know What It Means To Miss New Orleans" 3/7/06.

Radio Gets Wild – *www.radiogetswild.com/MReviews/op=show/rid=127.html* various tracks.

Radio Free Tunes – *www.radiofreetunes.com* – part of Artist Launch network – playing "Come On" and "Family of Love" for radio and podcast programs.

RhythmFlow Radio – *rhythmfl owradio.net* – playing various tracks.

Smooth Jazz Radio – *www.smoothjazz.com* – playing "Just In Time" – 2 spins daily.

Swissgroove Radio – *www.swissgroove.ch* – featuring CD on their "SmoothVibes."

Death Valley Radio – *http://deathvalleyradio.org/tracklists/track_list_404.html* – Family of Love

IQ Radio – *www.iqradio.net* – on various radio programs – also soon to be listing in their independent artists page feature section.

Ison Live Radio – *www.isonliveradio.com* – added 3/15/06 to rotation list – "Come On."

Radioio Acoustic – *www.radioio.com/acoustic* - playing "Family of Love" and "Just In Time."

Totally Acappella – playing various tracks – *www.totallyacappella.com.*

Solar Radio UK – no website was given – added "Come On" to their radio program 3/14/06.

Podcast

Deliberate Noise Podcast – *www.deliberatenoise.com.*

www.radiofreetunes.com playing "Come On" and "Family of Love" for radio/ podcast programs.

Lifeway – recording podcast 4/3.

Contest

ALL HIP HOP – *www.allhiphop.com* – 5 CD giveaways to correlate with interview.

FLASH MAGAZINE – *www.flashmag.de* – download contest being coordi-nated "Come On."

Detroitgospel.com/Chicagogospel.com – Take 6 loaded iPod give-away runs through 4/21.

Retailer/major promotions

AOL MUSIC: Listening party March 21–28 on AOLMusic listening page site.

AOL MUSIC: Download program "You Can Make It – Go on" begins 3/21.

AOL MUSIC: "Under the Influence" – brand new program associated with "Black Voices" – looking to book band for this in June during Blue Note shows.

AOL Sessions: pitch for June AOL Sessions date when band is back in NY 6/15–6/17.

iTunes Music Store – on newsletter for 3/21/06.

Real Networks/Rhapsody – exclusive track "What's Goin On" along with interview and celebrity playlist – *www.rhapsody.com/take6* will be up first week of April.

Sony Connect: Exclusive set taped in CA promoted in 3/28 newsletter.

YAHOO Music Unlimited – *music.yahoo.com/release/26227235* – NEWSLETTER TOP FEATURE ON 2/15/06.

Tourdate listings serviced to

www.upcoming.org
www.tribe.net
www.radicalendar.org
www.gotlocalmusic.com
www.musi-cal.com
www.mojam.com
www.jambase.com
www.citysearch.com
www.tollbooth.org
www.cabands.com
www.giglist.com/thescene/thescene.asp
www.myspace.com
www.dartradio.com
www.eventsetter.com
www.entertainment.lycos.com
www.inconcert.org
www.worldofgigs.com
www.calendarlive.com

Artist management contract form

This form is provided only as a guide to students of artist management for the recording industry. Although it gives a general view of contract components, an actual contract between an artist and an artist manager requires the guidance of their separate attorneys to ensure that the final contract is representative of the agreements and commitments made by both.

PERSONAL MANAGEMENT AGREEMENT

AGREEMENT made and entered into this _____ day of _____, by and between _____ ("Manager") located at _____ _____, on the one hand, and _____ p/k/a "_____" (individually and collectively referred to herein as "Artist") located at _____, on the other hand.

1. *Engagement and Term.*
    (a) Artist hereby appoints and engages Manager as Artist's sole and exclusive personal manager for a period of three (3) years commencing on the date hereof.
    (b) Manager shall have the option to extend the term of this Agreement for two (2) additional periods of two (2) years each. Such option periods shall commence automatically upon the expiration of the immediately preceding term of this Agreement unless Manager gives Artist notice to the contrary in writing not later than thirty (30) days prior to the end of any such term of Manager's desire to terminate this Agreement. Any reference to the term of this Agreement shall include the initial two (2) year period and any option periods.
2. *Manager's Services.*
    (a) During the term hereof, Manager agrees to use all reasonable efforts to promote, develop and advance Artist's professional career and to advise and counsel Artist with respect to all phases of Artist's career in the entertainment industry, including, without limitation, the following, to the extent applicable, to:

  (i) Advise and counsel in the selection of appropriate musical and other material and with regard to the adoption of the proper format for presentation of Artist's talents and the determination of proper style, mood, and setting in keeping with Artist's talents and best interests;

  (ii) Advise and counsel in any and all matters pertaining to employment, publicity, public relations and advertising;

  (iii) Analyze and comment upon Artists' music, physical appearance, wardrobe and other material;

  (iv) Advise, counsel and direct in the selection of artistic talent to assist, accompany or embellish Artist's presentation;

  (v) Advise and counsel with regard to general practices in the entertainment industry and with respect to such matters of which Manager may have knowledge concerning compensation and privileges for similar artistic values;

  (vi) Advise and counsel concerning the selection of theatrical agencies, and persons, firms and corporations who will counsel and advise Artist and seek and procure employment and engagements for Artist;

  (vii) Advise concerning the selection of persons such as brokers, accountants, attorneys, business managers, auditors and publicists to render services in connection with Artist's career;

  (viii) To the extent allowed by law, advise and counsel in connection with the negotiations of all agreements affecting Artist's career; and

  (ix) At Manager's sole discretion and if permitted by law, perform any other services customarily performed by a personal manager in the entertainment industry, provided that Manager shall not be obligated to seek, obtain or procure any employment or engagements for Artist.

(b) Manager shall be required to render all reasonable services which are called for by this Agreement as and when reasonably required. Manager shall not, however, be required to travel to meet with Artist at any particular place or places outside of Manager's primary place of business, except in Manager's discretion and subject to satisfactory arrangements for payment or reimbursement of the reasonable costs and expenses of such travel.

(c) Artist agrees that Artist's failure during the Term to continue to seek Manager's services shall not in any manner affect Artist's obligation to pay commissions and other monies to Manager as set forth herein.

3. *Non-Exclusivity/Affiliation.*

(a) Artist hereby acknowledges and agrees that Manager's services to Artist hereunder are non-exclusive and Manager shall at all times be free to perform the same or similar services for others in any field, regardless whether any such artist is in competition at any time with Artist, and engage in any other activities without limitation whatsoever as Manager in Manager's sole discretion shall decide. Additionally, from time to time during the Term, Manager and other persons, firms, or entities owned or controlled, in whole or in part, directly or indirectly by Manager or Manager's partners, shareholders, officers, directors, or employees, whether acting alone or in association with others, may have occasion to act in other capacities for Artist, including, without limitation, as record company, record producer, publisher and/or merchandiser. Such activity by Manager or such other person, firm or entity shall not be deemed a breach of this Agreement or of any obligations of Manager to Artist and shall not affect Manager's right to compensation hereunder, except as specifically provided herein. Notwithstanding the foregoing, Manager shall fully disclose to Artist all instances wherein Manager and/or other persons or entities (A) that are owned and/or controlled by Manager, (B) that own or control Manager, or (C) that are in common ownership and/or control with Manager ("Manager's Affiliates") proposes to participate in or take an ownership, income or other interest in a project or activity which creates an actual or potential conflict of interest with the business, financial or other interests of Artist. Except as expressly set forth herein, Manager shall not, without Artist's written consent, following such full disclosure, be entitled to Commission hereunder with Gross Earnings derived by Artist from: (i) any employment or agreement wherein Artist is employed by Manager (or Manager's Affiliates) in which Manager (or Manager's Affiliates) is acting as (a) the packager of the entertainment or amusement program in which Artist is so employed, or (b) the proprietor of any results or proceeds of Artist's services; (ii) from the sale, license or grant of any proprietary rights (e.g., music publishing rights) to Manager (or Manager's Affiliates); or (iii) the sale, license or grant of any literary or musical rights to Manager (or Manager's Affiliates).

(b) Notwithstanding the provisions in Paragraph 1(a) above, Artist acknowledges that Manager may associate with another personal manager ("Co-Manager") for the purpose of carrying out Manager's duties hereunder. Artist shall have the right to approve such Co-Manager provided that a Co-Manager shall be deemed pre-approved if at the time of Manager's affiliation such Co-Manager represents at least one (1) artist then signed to a so-called "major label." Notwithstanding the foregoing, Artist understands that Manager may choose to form and/or join a business entity, the business of which is, at least in part, the management of artists and Artist understands that an assignment of this Agreement to said entity is at Manager's sole discretion, and that consent from Artist will not be required.

4. *Manager's Authority and Power of Attorney.*

(a) Manager is authorized and empowered for Artist and in Artist's behalf, and in Manager's discretion, to do the following:

(i) Approve and permit any and all publicity and advertising;

(ii) Represent Artist in all dealings with any union;

(iii) Approve and permit the use of Artist's name, photograph, likeness, voice, sound effects, caricatures, literary, artistic and musical materials for purposes of advertising and publicity and in the promotion and advertising of any and all products and services or otherwise;

(iv) Prepare, negotiate, consummate, sign, and execute for Artist, in Artist's name and/or in Artist's behalf, any and all agreements, documents and contracts for Artist's services, talents, commercial endorsements and/or artistic, literary and musical materials, including, without limitation, television and live performances and personal appearances;

(v) Engage, as well as discharge and/or direct for Artist and in Artist's name, accountants, business managers, auditors, theatrical agents, publicists, attorneys, booking agents and employment agencies, as well as other persons, firms and corporations who may be retained in connection with Artist's business and/or financial affairs, and/or to obtain contracts, engagements, or employment for Artist or otherwise act in connection with Artist's career; provided however, that Artist and Manager shall mutually agree upon the selection of a business manager ("Business Manager"), which Business Manager shall be engaged at Artist's sole expense. Artist shall consult with Manager prior to its engagement or dismissal of any of the foregoing parties, provided that a failure to do so under exigent circumstances, where it is not possible to consult Manager, shall not be deemed a breach hereof. Notwithstanding the foregoing, Manager shall neither engage nor discharge attorneys or business managers on Artist's behalf without Artist's prior consent.

(vi) Prior to the engagement of a Business Manager, collect and receive monies on behalf of Artist, to endorse Artist's name upon and deposit or cash any an all checks payable to Artist for Artist's services, talents, and music materials, and to retain thereof all sums owing to Manager and Artist shall instruct any agency or Artists' manager engaged by Artist to remit to Manager all monies that may become due Artist and may be received by such agency or Artists' manager; and

(vii) Subsequent to the engagement of a Business Manager, irrevocably direct and authorize in Artist's name or on Artist's behalf all persons and companies owing monies to Artist to deliver gross earnings directly to Business Manager and direct Business Manager to compensate parties set forth above (including, without limitation, Manager's commissions) out of Artist's funds (Artist hereby expressly acknowledges that Artist is solely responsible for their compensation and that Artist shall indemnify and hold Manager harmless with respect thereto) and to audit and examine the books and records of parties with whom Artist has contracted.

(b) Artist hereby grants to Manager the full power, right and authority as Artist's true lawful attorney-in-fact to do any and all of the foregoing acts as Manager shall, in Manager's sole discretion, deem advisable in the same manner as Artist could if personally present, and Artist hereby ratifies and confirms any and all acts that Manager shall perform or cause to be performed in accordance with the foregoing, which power shall be irrevocable during the Term hereof.

5. *Offers of Employment.* Artist shall advise Manager of all offers of employment submitted to Artist and will refer any inquiries concerning Artist's services to Manager, for Manager's advice and approval and in order that Manager may determine whether the same are compatible with and in the best interest of Artist's career.

6. *Expenses.*

(a) Manager is not required to make any loans or advances to Artist or for Artist's account, but in the event Manager does so, Artist shall remain primarily liable to repay such loans and/or advances, as applicable, Artist hereby authorizes Manager to either deduct the amount of any such loan (s) or advance(s) from any sums received by Manager on Artist's behalf and/or shall cause Business Manager to repay to Manager the amount of such loan(s) and/or advance(s), as applicable, promptly or at such other time as may be agreed to in writing between Artist and Manager. The foregoing authorization is coupled with an interest and shall be irrevocable during the Term.

(b) (i) Artist shall be solely responsible for payment of all booking agencies, fees, union dues, publicity costs, promotional or exploitation costs, traveling expenses and/or wardrobe expenses and reasonable expenses arising from the performance by Manager of services hereunder.

(ii) Artist hereby irrevocably authorizes Manager to incur the following expenses, among others, during the Term: long-distance telephone charges; photocopying charges; postage charges; facsimile and telex transmission charges; messenger and courier charges; accountants' fees and costs; video and tape duplication charges; air and ground transportation and lodging and meals and other living expenses while traveling; promotion and publicity expenses; and any other disbursements attributable to Artist. Automobile travel outside Tacoma, Washington shall be reimbursed at the then-current IRS standard rate per mile or for the cost of a rental car in lieu of such mileage. Manager's class of air transportation on a particular trip shall be the same as Artist's, provided that Manager shall not be required to travel to meet with Artist at any particular place.

(iii) Notwithstanding the above,

(A) Manager shall not incur any expense for which Artist would be responsible nor shall Manager have the right to reimbursement as to any single expenditure exceeding One Thousand Dollars ($1,000) or a group of related expenses which exceed Two Thousand Five Hundred Dollars ($2500) per month without Artist's prior consent in each instance.

(B) Expenses which concern both Artist and one or more other clients of Manager (including, without limitation, travel-related expenses) shall be prorated in an appropriate manner.

(C) Artist shall cause Business Manager to reimburse Manager for expenses incurred by Manager pursuant to this paragraph 6 within fifteen (15) days after receipt by Business Manager of itemized documentation of such expenses; provided, however, that Artist shall not be required to cause Business Manager to make expense reimbursement payments to Manager more frequently than once per month. Artist hereby authorizes and empowers Manager to deduct and retain the amounts of such reimbursable costs and expenses from any sums Manager may receive for Artist's account pursuant to this Agreement.

7. *Artist's Duties.* Artist agrees at all times to pursue diligently and faithfully Artist's career on a full time basis, to the best of Artist's ability, to do all things necessary and desirable to promote such career and maximize earnings therefrom and to refrain from unlawful or offensive conduct. Artist agrees to study, practice, train and rehearse as Manager may direct and to attend and perform at any rehearsal, practice or training session which Manager may arrange. Artist shall at all times utilize proper theatrical or other employment agencies, as applicable, to obtain engagements and employment for Artist except as may otherwise be permitted by law. Artist shall endeavor to discuss all offers of employment with Manager and shall refer any inquiries or offers concerning Artist's services to Manager as well as to Artist's licensed talent agent. Artist shall irrevocably instruct any theatrical or other employment agent engaged by Artist to remit directly to Business Manager all monies that may become due to Artist that are received by such agent. Further, because careful planning and coordination are required for the performance of Manager's duties and obligations hereunder, and to maximize Artist's career development and income, Artist will not, during the Term, take part in any performance, engagement, or act, nor perform, sing or otherwise display or exploit Artist's talents in any manner, whatsoever, in any place whatsoever, except after consultation with Manager, and Artist will not allow Artist's name to be used in any professional or commercial enterprise, whatsoever, except after consultation with Manager. Artist shall not enter into any agreement or commitment which shall in any manner interfere with Manager's carrying out the terms and conditions of this Agreement. Without limiting the foregoing, during the Term, Artist shall not, without Manager's prior written consent, engage any other person, firm or corporation to render any

services of the kind required of Manager hereunder or which Manager is permitted to perform hereunder.

8. Excluded Services. Artist understands and hereby acknowledges that Manager is not an employment agent, theatrical agent or business manager, and that, except to the extent permitted by law, Manager has not offered, attempted or promised to obtain employment or engagements for Artist, and that, except to the extent permitted by law, Manager is not permitted, obligated, authorized or expected to do so.

9. *Relationship of Parties.* This Agreement shall not be construed to create a partnership between Manager and Artist. It is specifically understood that Manager is acting hereunder as an independent contractor and Manager may appoint or engage other persons, firms and corporations, in Manager's discretion, to perform services which, Manager has agreed to perform hereunder; provided, however, Manager agrees to act as the personal manager for Artist during the Term hereof. During the Term, Manager shall have the exclusive right to advertise and publicize Manager as Artist's personal manager.

10. *Commissions.*

    (a) In consideration of Manager's services hereunder, Manager is and shall be entitled to receive from Artist (who agrees to pay to Manager, as and when received by Artist) or directly from third parties pursuant to the provisions hereof, for the Term hereof and thereafter as set forth in this Agreement, a sum equal to twenty percent (20%) ("Commission") of any and all gross earnings (as hereafter defined) which Artist may receive as a result of Artist's activities in and throughout the entertainment, theatrical, motion picture, television, music, and recording industries and all other areas in the entertainment industry in which Artist's artistic talents are developed and exploited, including any and all sums resulting from the use of Artist's artistic talents and the results and proceeds thereof. Without in any manner limiting the foregoing, the matters upon which Manager's compensation shall be computed shall include any and all of Artist's activities in connection with: motion pictures; television; radio; music; literary; theatrical engagements; personal appearances; public appearances in places of amusement and entertainment; records and recordings; publications; and the use of Artist's name; voice; likeness and talents for purposes of merchandising; advertising; endorsements and trade.

    (b) The term "gross earnings" as used herein shall mean and include any and all gross monies or other considerations which Artist, or any designee, assignee, legatee, executor, heir, administrator, trustee, conservator or successor of Artist, attorney, Business Manager or any other person or company, including, without limitation, Manager, acting for or on behalf, or for the benefit, of Artist (collectively and individually the "Artist Parties"), may receive or which may be payable to Artist or the Artist Parties or credited to Artist's account or any account of the Artist Parties (without any exclusion or deduction, including, without limitation, without deductions for taxes (whether withheld or payable) of any kind, union dues or any other expenses, fees, costs or obligations) as a result of Artist's activities, services or exploitation of rights in and throughout the entertainment and amusement industries throughout the universe, whether as, without limitation, an actor, performer, choreographer, dancer, consultant, author, lyricist, record producer, engineer, songwriter, composer, publisher, commercial endorser, product endorser, sponsored or tie-in artist, singer, musician, artist, technician, film producer, director, supervisor, executive, consultant or proprietor, including, without limitation, any and all gross monies or other considerations resulting from the use of Artist's artistic talents (whether as a solo artist, a member of a group or otherwise), the name, likeness or other identifying features or characteristics of Artist and the results and proceeds thereof and the sale, lease or other disposition of musical, literary, dramatic or other artistic material which Artist may create, compose or collaborate, directly or indirectly, in whole or in part, in all fields, and any act, unit or package show of which Artist may be the owner or part owner, directly or indirectly, or otherwise have any interest in. Without in any manner limiting the foregoing, the Commission shall be payable with respect to Artist's gross monies or other considerations in respect of any and all of Artist's activities throughout the universe ("Employment Activities") in connection with, without limitation, motion pictures, television, video, radio, publishing, literary, artistic and musical material, theatrical engagements, personal appearances, public appearances in places of amusement or entertainment, sponsorship, endorsements, records and recordings, publications and the use of Artist's name, likeness, identity and talents for purposes of merchandising, advertising, commercial exploitation, trade and otherwise. The term "gross monies or other considerations" shall include, without limitation, salaries, earnings, fees, royalties or advances against royalties, gifts, bonuses, shares of profit, shares of stock, partnership interests, participations and percentages and the total amount paid for a package television or radio program (live or recorded), motion picture or other entertainment or amusement package, earned or received directly or indirectly by, or credited to any account of, Artist or Artist's heirs, executors, administrators or assigns, or by any other person, firm or corporation on Artist's behalf or for Artist's benefit. In the event that Artist receives, as all or part of Artist's compensation for activities covered hereunder, stock or the right to buy stock in any corporation, or that Artist becomes the packager or owner of all or part of an entertainment property, whether as individual proprietor, stockholder, partner, joint venturer or otherwise, Manager's commission shall apply to Artist's said stock, right to buy stock, individual proprietorship, partnership, joint venture, or other form of interest, and Manager shall be entitled to Manager's percentage share thereof. Should Artist be required to make any payment for such interest, Manager will pay Manager's percentage share of such payment, unless Manager elects not to participate therein. In the event that any corporation, partnership, trust, joint venture, association or proprietorship or other business entity (hereinafter "firms") in which Artist, or any member of Artist's family, has a direct or indirect interest, shall receive any compensation for permitting or contracting for the use of Artist's services, name, likeness or endorsement, then such compensation shall be deemed to be "gross earnings" received by Artist for purposes of this Agreement.

(c) The Commission shall be paid to Manager in perpetuity as and when gross earnings in respect of Employment Activities are received or credited to any of the payees specified in paragraph 10 above (or paragraph 10(h) below), without any geographical or temporal limitation, except as specifically set forth herein, pursuant to:

(i) any and all agreements, engagements and commitments now in existence;

(ii) any and all agreements, engagements and commitments entered into, substantially negotiated, or commenced during the Term, provided that only the following master recordings and musical compositions are commissionable:

(A) master recordings featuring Artist which were recorded before the Term but released during the Term and those recorded during the Term;

(B) musical compositions written in whole or in part during the Term and those musical compositions written before the Term which are recorded by Artist during the Term. Notwithstanding the foregoing, if Manager obtains a third-party cover recording of a musical composition written before the Term which has not been recorded by Artist, or if Manager assists and/or counsels Artist in the entry (or substantial negotiation of the terms) of a third party agreement with respect to such musical compositions written before the Term, then Manager shall commission such composition(s).

(C) Notwithstanding anything to the contrary in paragraphs 10(c)(ii) (A) and (B), if an agreement, engagement or commitment for Artist's services in connection with a master recording and/or a musical composition is entered into, substantially negotiated or commenced during the Term, but such recording is not actually recorded until after the Term and/or such composition is not actually written and/or recorded until after the Term, Manager shall nonetheless be entitled to commission such master recording and/or composition. The foregoing shall apply, without limitation, to projects such as a motion picture or television soundtrack and/or soundtrack album, the composing and/or recording of theme songs, and similar engagements.

   (iii) any and all extensions, additions, substitutions, renewals, replacements, modifications and amendments, without any limitation of time, of all such agreements, engagements and commitments referred to in subparagraphs (i) and (ii) hereof;

   (iv) any and all copyrights and publishing or other rights in any entertainment or amusement properties owned or acquired (by assignment, license or other means), in whole or in part, by Artist or the Artist Parties prior to or during the Term (an agreement to acquire same will be commissionable if entered into or substantially negotiated prior to or during the Term), shall be commissionable hereunder; and

   (v) any and all judgments, awards, settlements, payments, damages and proceeds relating to any suits, claims, actions, proceedings or arbitration proceedings arising out of any alleged or actual breach, nonperformance or infringement by others of any of the agreements, engagements, commitments, other agreements or rights referred to in subparagraphs (i), (ii), (iii) and (iv) hereof, all of which regardless of when entered into, when performed and when effective.

(d) Notwithstanding the foregoing to the contrary, Manager shall receive a sum equal to the percentages set forth below with respect to all gross monies or other considerations earned by Artist as a result of Artist's activities in and throughout the entertainment, theatrical, motion picture, television, music, and recording industries and all other areas in the entertainment industry in which Artist's artistic talents are developed and exploited, including any and all sums resulting from the use of Artist's artistic talents and the results and proceeds thereof, following the expiration of the Term hereof in connection with the engagements, contracts and agreements which are entered into prior to or during the Term of this Agreement or substantially negotiated during the Term hereof and entered into thereafter, and any renewals or extensions of those engagements, contracts or agreements in which Artist may earn or receive Gross Earnings:

   (i) Fifteen percent (15%) during the first and second year following the expiration of the Term hereof;

   (ii) Ten percent (10%) during the third and fourth year following the expiration of the Term hereof;

   (iii) Five percent (5%) during the fifth and sixth year following the expiration of the Term hereof;

   (iv) Nothing thereafter.

(e) Notwithstanding anything to the contrary, the following amounts shall be deducted from, and shall not be included in, the gross earnings upon which Manager's Commission is based: (i) direct recording costs (other than advances, production and session fees Artist retains for Artist's personal use), including advances and royalties to third party producers and engineers; (ii) direct production costs (other than advances Artist retains for Artist's personal use) for music video and any other audiovisual productions; (iii) "deficit tour support" (as that term is understood in the music industry) and any other amounts paid to Artist or on Artist's behalf which are used to offset any deficit in connection with Artist's personal appearance(s) (in clarification of the foregoing, if Artist receives an engagement fee of $1,000 but receives tour support of $2,000 to offset expenses, Manager shall nonetheless commission the $1,000 engagement fee); (iv) direct sound and light costs in connection with Artist's appearances actually paid to unaffiliated third parties (not to exceed 25% of the gross revenue derived from such production), provided that if a promoter allocates an amount as reimbursement or payment for sound and lights but a lesser amount is actually paid to unaffiliated third parties, the difference will be included in gross earnings; (v) opening and support act costs, if paid by Artist; (vi) payments (including advances and royalties) due to third party songwriters and publishers because of such songwriters' creative contributions to a particular composition; (vii) expense reimbursements for costs expended by or on behalf of Artist; (viii) bona fide gifts; and (ix) personal injury and punitive damage awards.

(f) The interest and compensation set forth in this Agreement which shall be paid to Manager shall be a continuing interest, and shall not be revocable at Artist's pleasure. It is intended by Artist to create an agency coupled with an enforceable interest therein, and the appointment and engagement of Manager and Manager's right to receive the Commission are the inducements for Manager's entering into this Agreement.

(g) The term "substantially negotiated" as used herein shall mean that negotiations have proceeded to a point where specific terms of the agreement have been discussed in detail, it being understood that such negotiations will be more than a mere solicitation of interest, but need not have proceeded to the point where an offer has been made.

(h) Artist shall cause any corporation, partnership, trust, joint venture, association, proprietorship or other business entity in which Artist or any member of Artist's family has any direct or indirect interest of any nature or sort, or which is directly or indirectly controlled by Artist or under the common control of Artist and others (hereinafter "firms") to enter into an agreement with Manager on the same terms and conditions as this Agreement, and Artist agrees that all gross monies or other considerations directly or indirectly earned or received by such firms in connection with Artist's activities in the entertainment industries shall be subject to Manager's commission hereunder. However, Manager shall not be entitled to any double commissions by reason of this paragraph 10(h).

11. *Accounting.* Artist shall cause Business Manager to prepare and maintain books and records relating to monies received by Business Manager on Artist's behalf. Within ten (10) days following the 1st and 15th days of each month during which Gross Earnings are received, Artist shall render (or cause the Business Manager to render) to Manager a statement showing the amount

of Gross Earnings received, the source and the nature thereof, and the amount of the money due and payable to Manager in commissions and expenses for such period, accompanied by payment of any monies shown to be due. Artist (or Artist's Business Manager) shall also furnish Manager with copies of all statements rendered to Artist by third parties in connection with the payment of Gross Earnings on which commissions are not payable to Manager. Artist and Manager shall each have the right to inspect Business Manager's books and records with regard to monies received on Artist's behalf upon thirty (30) days' notice during reasonable business hours, twice per calendar year. If Artist receives monies commissionable hereunder directly, Artist shall remit such monies to Business Manager within five (5) days of receipt. Manager shall have the right to inspect Artist's books and records with respect to Artist's gross earnings. Notwithstanding anything to the contrary in this Agreement, if and during such times as no Business Manager is engaged on Artist's behalf, then Manager shall hereby be authorized, but not required, to assume the rights and obligations of Business Manager hereunder until such time as a new Business Manager has been engaged by Artist.

12. *Assignment.* Manager shall have the right to assign this Agreement to any third party associated or affiliated with Manager or in which any of the proprietors of Manager have a direct or indirect interest or which owns or acquires a portion of Manager's assets. This Agreement is personal to Artist, and Artist agrees not to assign this Agreement or any portion thereof, any such assignment to be ineffective and null and void.

13. *Representations and Warranties.* Artist warrants, represents and agrees that Artist is not a minor; that Artist is under no disability, restriction or prohibition with respect to Artist's right to execute this Agreement and perform its terms and conditions; is free to enter into this Agreement; is not at this time (and will not so become during the Term) under contract to any other personal manager; will name Manager and _____ as additional insureds on all liability insurance policies (including tour-related insurance) and errors and omissions policies on which Artist is named during the Term; and has not heretofore made and will not hereafter enter into or accept any engagement, commitment or agreement with any person, firm or corporation which will, can or may interfere with the full and faithful performance by Artist of the covenants, terms and conditions of this Agreement to be performed by Artist or interfere with Manager's full enjoyment of Manager's rights and privileges hereunder. Artist agrees to exert Artist's best efforts to further Artist's career during the Term of this Agreement, and to cooperate with Manager to the fullest extent in the interest of promoting Artist's career.

14. *Indemnity.* In the event that Artist does not fulfill or cause to be fulfilled any agreement or obligation undertaken by Artist, Artist agrees to indemnify and hold Manager harmless from any claims, demands, actions, judgments and awards against Manager by third parties in connection with such non-fulfillment.

It is agreed that Manager will not be held liable or responsible for any breach of contract or act or omission on the part of any person, firm or corporation with respect to any engagement or agreement concerning Artist's services. Artist shall at all times defend (subject to Manager's option to elect to defend itself), indemnify and hold harmless Manager, Manager's Affiliates and Manager's officers, partners and employees from and against any and all claims, damages, liabilities, costs, expenses (including, without limitation, attorneys' fees and costs) and settlements, judgments and awards arising out of any alleged or actual breach by Artist of any warranty, representation or agreement made by Artist herein or with regard to any third party. Artist shall reimburse Manager and Manager's Affiliates and Manager's officers, partners and employees, on demand, for any payment made at any time after the date hereof in respect of any liability or claim in respect of which Manager, Manager's Affiliates or Manager's officers, partners or employees are entitled to be indemnified. In the event Artist fails to so reimburse Manager, Manager's Affiliates or Manager's officers, partners or employees, Manager may, in addition to its other rights, direct Business Manager to remit to Manager such amounts from any and all gross earnings received by Business Manager hereunder.

15. *Remedies.*

(a) Artist hereby acknowledges and agrees that Manager's right to represent Artist as Artist's sole and exclusive personal manager and Artist's obligation to use Manager solely and exclusively in such capacity are unique, irreplaceable and extraordinary rights and obligations and that any breach or threatened breach by Artist thereof shall be material and shall cause Manager immediate and unavoidable damages which cannot be adequately compensated for by a money judgment. Accordingly, Artist agrees that, in addition to all other forms of relief and all other remedies available to Manager in the event of any such breach or threatened breach by Artist, Manager shall be entitled to seek injunctive or other equitable relief against Artist to enforce Manager's rights hereunder.

(b) If at any time Artist fails, for any reason whatsoever, to fulfill or perform any obligation assumed by Artist, then, without limiting Manager's rights, Manager shall, without limitation of Manager's other rights, have the right, exercisable at any time by notice to Artist, to extend the expiration date of the then-current period of the Term. Such extension shall continue until Artist has cured fully such failure and the then-current period of the Term shall be extended for a period of time equal to the duration of any such failure, subject to such limitation, if any, as are imposed by law.

(c) In the event of any dispute arising under or relating to this Agreement which results in litigation, the non-prevailing party shall, in addition to any damages or other relief awarded, pay the prevailing party reasonable costs of litigation, including, but not limited to, reasonable attorney's fees.

16. *Notices.* All notices hereunder shall be delivered by personal service, overnight commercial courier, or sent by certified or registered mail, return receipt requested, postage prepaid, and if to Manager shall be sent to the address set forth on the first page hereof, with a copy to _____ and, if to Artist, shall be sent to Artist's address set forth on the first page hereof, unless the parties notify each other as provided herein that notices should be sent to a different address. The date of mailing or in the case of personal service or courier, the date of delivery, shall be deemed the date of the giving of said notice except notice of change of address which shall be effective only upon actual receipt.

17. *Life Insurance.* Manager shall have the right during the Term to obtain life insurance on Artist's life at Manager's cost, with Manager being the sole beneficiary thereof and for an amount to be determined by Manager in Manager's sole discretion. Artist shall cooperate fully in connection with obtaining such insurance, and Artist shall submit to a physical examination and complete promptly all documents necessary or desirable for such insurance, Artist hereby acknowledges that neither Artist nor Artist's estate shall have any right to claim the benefits of any such policy obtained by Manager.

18. *Artist Membership.*

    (a) Artist now comprises the group professionally known as "_____". In the event that one or more members of Artist leave the group as presently constituted, Manager shall have the option to terminate this Agreement upon thirty (30) days notice to Artist. Manager shall also have the option to continue to manage any leaving member(s) of the group upon thirty (30) days written notice to such leaving member(s) under the same terms and conditions herein, but in no event shall the term of the Agreement with regard to such leaving member exceed the original Term of this Agreement. Notwithstanding anything to the contrary, if a leaving member joins an existing group which is then subject to a personal management agreement, it shall not be deemed a breach of this Agreement if Manager does not become the manager or co-manager of such existing group.

    (b) If Artist engages or agrees to have additional persons perform with Artist as part of "_____", which Artist shall only do following consultation with Manager, then such person shall also be required to agree to be bound to this Agreement and shall be required to enter into an agreement with Manager on the same terms and conditions as the other members comprising Artist.

    (c) In the event of death of Artist or a member of Artist, this Agreement shall continue in full force and effect and Manager shall provide management services to the deceased Artist's estate (or the estate of the deceased member of Artist).

19. *Approvals.* For all instances where approvals must be obtained from Artist, Manager may obtain such approval by communicating the request to all members of Artist via an appropriate method of communication based on the circumstances of Artist' location and schedule at the relevant time and the

exigencies of the needed approval (e.g., email, Artist's cellular telephone, facsimile or personal delivery) and Manager's receipt of a response from the applicable Artist delegate (which will be specified in writing by Artist) shall be deemed to be an authorized response on behalf of Artist to such request. For purposes of the foregoing, the Artist delegate from commencement hereof until the end of the _____ quarter of 2005 (_____, 2005) is hereby designated as _____ . It is understood that such Artist delegate shall communicate internally with the other members of Artist prior to granting or denying any such approvals on behalf of Artist. Artist agrees to cooperate in good faith to reply to Manager's requests for approval.

20. *Legal Representation.* Artist represents that Artist has been represented by independent counsel or have had the unrestricted opportunity to be represented by independent legal counsel of Artist's choice for purposes of advising Artist in connection with the preparation, negotiation and execution of this Agreement. If Artist has not been represented by independent legal counsel of Artist's choice in connection with this Agreement, Artist acknowledges and agrees that Manager has encouraged Artist to obtain such representation and Artist's failure to be represented by independent legal counsel in connection with this Agreement was determined solely by Artist.

21. *Miscellaneous Provisions.*

    (a) This Agreement shall be deemed to have been executed in and shall be construed in accordance with the laws of the State of _____ . Any controversy arising under this Agreement shall be adjudicated under the jurisdiction of a competent court within _____ County, _____ . If any provision hereof shall for any reason be illegal or unenforceable, such unenforceability shall not affect the validity of the remaining portion and provisions hereof. If this Agreement is, for any reason, invalid, illegal, or unenforceable, Artist agrees that Manager shall nevertheless be entitled to the reasonable value of Manager's services and shall be entitled to retain all compensation paid to Manager hereunder as the reasonable value of such services.

    (b) Except as expressly provided in this Agreement to the contrary, this Agreement is made only for the benefit of Manager and Artist. No other person, firm or corporation shall have or acquire any rights hereunder.

    (c) Neither party shall be deemed to be in breach of any of such party's obligations hereunder unless and until the other party shall have given the allegedly breaching party specific written notice by certified or registered mail, return receipt requested, of the nature of such breach and the allegedly breaching party shall have failed to cure such breach within thirty (30) days after such party's receipt of such written notice if such breach is capable reasonably of being cured fully within such thirty (30) day period, or if such breach is not capable reasonably of being cured fully within such thirty (30) day period, if such party commences to cure such breach within such thirty (30) day period and proceeds with reasonable diligence to cure such breach.

(d) This Agreement is the only agreement of the parties and there is no other agreement, oral or written, between the parties relating to the subject matter hereof.

IN WITNESS WHEREOF, the parties hereto have executed this Agreement as of the date first above written.

MANAGER: **ARTIST:**

By: _____   _____

                                         SSN: _____

                                         Date of Birth: _____

Partnership agreement for members of a band

This form is provided only as a guide to students of artist management for the recording industry. While it gives a general view of components of a partnership agreement for members of a band or a group, the preparation and execution of an actual agreement requires the guidance of their separate attorneys to assure that the final version is representative of the agreements and commitments made by each.

General partnership agreement of

THIS AGREEMENT is entered into as of the _____ day of _____, _____, by and among _____. Hereinafter, the parties shall sometimes be referred to collectively as "Partners" and individually as "Partner."

Recitals

A. The Partners have been operating a business (the "Prior Partnership") under an oral partnership agreement (the "Prior Agreement").

B. The Partners desire to enter into a written agreement concerning the business of the Partners, which agreement shall supersede and cancel all prior understandings and agreements, oral or written, concerning such business, including the Prior Agreement.

NOW, THEREFORE, in consideration of the mutual promises contained herein, the parties hereby agree as follows:

Article I **Formation of general partnership**

1.1 *Type of Business*. The Partners hereby agree to establish a general partnership (the "Partnership") which, in accordance with this agreement, shall engage in the business of utilizing and commercially exploiting their collective talents and personalities as the recording and performing group professionally known as "_____" in all media of public entertainment throughout the world, including phonograph recordings, motion pictures, television, radio, public appearances and performances, advertising, merchandising, literary and dramatic endeavors, publications, and all other similar or related activities, but excluding songwriting and music publishing.

1.2 *Name of Partnership*. The name of the Partnership shall be "_____, a Tennessee general partnership" and the Partners shall perform, as a recording group (sometimes hereinafter referred to as the "Group") under the professional name, trademark and service mark "_____" (sometimes hereinafter referred to as the "Group Name").

1.3 *Term of Partnership*. The Partnership shall continue until dissolved in accordance with this agreement.

1.4 *Place of Business*. The principal place of business of the Partnership shall be Nashville, Tennessee, or such other place or places as may be designated by Partnership Vote (as such term is defined in paragraph 6.1 below).

Article II Partnership capital and loans

2.1 *Initial Capital*. The initial capital of the Partnership shall consist of all assets and obligations of the Prior Partnership, including the following:
  (a) all right, title, and interest in and to the Group Name;
  (b) all trademarks, service marks, logos and goodwill of the Prior Partnership;
  (c) that certain recording agreement by and between the Partners and _____, dated _____ (the "Recording Agreement"); and
  (d) all equipment and other assets owned by the Prior Partnership, including without limitation, that equipment and other assets set forth in Exhibit "A" hereto.

2.2 *Capital Accounts*.
  (a) Capital accounts will be maintained for the Partners in accordance with Treasury Regulation Section 1.704-1(b)(2)(iv). The Partners shall decide by Partnership Vote (hereinafter defined) whether to make any elections or adopt any conventions relating to the foregoing.
  (b) The Partners hereby agree that for purposes of valuing the Partners' capital accounts, the assets specified in paragraphs 2.1(a), (b) and (c) shall be deemed to have zero value.

2.3 *Additional Contributions to Capital*. No Partner shall be required to make additional contributions to the capital of the Partnership. Any voluntary contributions to the capital of the Partnership must be authorized by Partnership Vote. Except as otherwise determined by Partnership Vote, any such voluntary contributions shall not alter the profit and loss allocations set forth in Article IV.

2.4 *Withdrawal of Capital*. No Partner shall withdraw any portion of the capital of the Partnership without the unanimous approval of the Partners. Any permitted withdrawal of capital by a Partner shall reduce such Partner's capital account by the amount of such withdrawal.

2.5 *Interest on Capital.* No Partner shall be entitled to interest on his share of the capital of the Partnership; provided, however, that this paragraph 2.5 shall not be construed to diminish any Partner's share of interest earned by the Partnership on Partnership funds.

2.6 *Loans to Partnership.* No Partner shall loan or advance money to the Partnership without approval by Partnership Vote. Any loan by a Partner to the Partnership shall be separately recorded in the Partnership books as a loan to the Partnership, shall bear interest at a rate determined by Partnership Vote, and shall be evidenced by a promissory note delivered to the lending Partner and executed in the name of the Partnership by one or more non-lending Partners.

Article III **Accountings**

3.1 *Method of Accounting.* The accounting books of the Partnership shall be kept on a cash basis.

3.2 *Fiscal Year.* The fiscal year of the Partnership shall be the calendar year.

3.3 *Accountings.* As soon as reasonably practicable, but no later than ninety (90) days after the close of each fiscal year, a full and accurate accounting shall be made of the affairs of the Partnership as of the close of such fiscal year. As part of such accounting, the Net Profits and Net Losses (as such terms are hereinafter defined) sustained by the Partnership during such fiscal year shall be ascertained, and credited or debited, as the case may be, to the respective Partners' capital accounts. As a further part of such accounting, the accountant for the Partnership designated from time to time by Partnership Vote (hereinafter referred to as the "Accountant") shall furnish to each Partner a copy of a balance sheet of the Partnership as of the last day of such fiscal year, a statement of income or loss of the Partnership for such fiscal year, a statement of allocations pursuant to this agreement for such fiscal year, and a statement of tax allocations for such fiscal year pursuant to the Code.

3.4 *Books of Account.* Complete and accurate books of all transactions of the Partnership shall be maintained at the principal office of the Accountant or at such other place or places that may be designated by Partnership Vote. Each Partner shall, at all reasonable times, have access to the Partnership's books and may inspect and copy such books, provided that any such inspection is made in good faith and without intent to damage the Partnership or any of the Partners.

3.5 *Bank Accounts.* All funds of the Partnership shall be deposited in accounts in the name of the Partnership at such bank or banks that shall be designated by Partnership Vote. All withdrawals therefrom shall be made upon checks or withdrawal slips signed by such person or persons as from time to time shall be designated in writing by Partnership Vote. Until otherwise designated in writing by Partnership Vote, all checks or withdrawal slips of the Partnership shall be signed by the Accountant.

Article IV **Profits, losses, distributions and expenses**

4.1 *Allocation of Profits and Losses.* Net Profits and Net Losses (as such terms are defined in paragraph 4.2 below) of the Partnership shall be allocated equally among the Partners.

4.2 *"Net Profits" and "Net Losses."* The terms "Net Profits" and "Net Losses" as used herein shall mean and refer to net profits and net losses for federal income tax purposes with such adjustments as required to comply with Treasury Regulation Section 1.704-1(b)(2)(iv). An allocation of Net Profits and Net Losses shall be reflected in each Partner's capital account.

4.3 *Distributions and Draws.*

    (a) (i) Promptly following the rendering of financial statements for each fiscal year of the Partnership, and after consulting with the Accountant, the Partnership shall distribute to the Partners any cash not reasonably necessary for the operation of the Partnership business. The Partnership shall make additional distributions to the Partners on other dates and in amounts to be determined by Partnership Vote.

        (ii) Unless otherwise agreed upon by Partnership Vote, all distributions (other than draws pursuant to paragraph 4.3(b) below) shall be allocated equally among the Partners.

    (b) The Partners shall be entitled to withdraw monthly from the funds of the Partnership, or on such other periodic or non-periodic basis as shall be determined from time to time by Partnership Vote, a sum as shall be determined by Partnership Vote after due consultation with the Accountant (taking into consideration, among other things, the advisability of maintaining a reasonable reserve for future anticipated expenses), and which sum, unless otherwise agreed by the unanimous vote of all the Partners, shall be divided equally among all the Partners.

4.4 *Salaries and Guaranteed Compensation.*

    (a) Except as otherwise specified in paragraph 4.4(b), and except as otherwise decided by Partnership Vote, no Partner shall be entitled to any salary or guaranteed compensation.

    (b) While performing during any recording session of the Partnership, the Partners shall be treated as royalty artists, and shall receive, as salary, single scale for one session (as such term is commonly used in the music industry) for each recording track they perform on. The amount of any such salary shall constitute a Partnership expense.

4.5 *Reimbursement of Expenses.* No Partner shall be reimbursed for payments made or liabilities incurred in the conduct of Partnership business or for the preservation of Partnership property unless such reimbursement is approved by Partnership Vote; provided, however, that the Accountant may, without approval by Partnership Vote, reimburse a Partner from Partnership funds for any payment or liability incurred in the conduct of Partnership business or for the preservation of Partnership property to the extent such payment or liability for any such transaction does not exceed $ _____.

Article V Duties of partners, rights of partnership, outside activities and publishing participation

5.1 *Services Furnished and Rights Granted to the Partnership.*

   (a) *Services Performed.* Each Partner shall furnish exclusively to the Partnership, throughout the world, such Partner's services in all media of public entertainment described in paragraph 1.1 above (except with respect to Outside Activities [as defined and permitted in paragraph 5.2(b) below]) and the results and proceeds thereof, as are necessary to permit the Partnership to carry out its purpose as described in said paragraph 1.1. Without limiting the generality of the foregoing, each Partner shall (i) render all services necessary to enable the Group to satisfy its obligations to perform concerts and other "live" performances, including without limitation, attending all rehearsals scheduled by the Partnership, and (ii) furnish to the Partnership such Partner's services as a recording artist, which services shall be sufficient to enable the Group to completely satisfy its obligations under the Current Recording Agreement or any successor recording agreement(s) entered into by the Partnership after the expiration or termination of the Current Recording Agreement (individually and collectively "Successor Recording Agreement") (hereinafter the Current Recording Agreement, any extensions or renewals thereof, and any Successor Recording Agreement shall be sometimes individually and collectively referred to as the "Recording Agreement"). Each Partner shall refrain from taking any action, including any Outside Activities, that would place the Partnership in breach of the Recording Agreement or any other agreement entered into by the Partnership in accordance with this agreement (hereinafter any such agreement entered into by the Partnership, including the Recording Agreement, shall sometimes be referred to as an "Approved Agreement"), and shall otherwise render all services necessary to enable the Partnership to satisfy its obligations under each Approved Agreement.

   (b) *Grant of Rights.* Each Partner hereby grants to the Partnership all rights in and to the results and proceeds of such Partner's services performed under this agreement, and all rights in and to such Partner's name, photograph, likeness, voice, and biographical materials for use in the exploitation of products and services of the Partnership. Without limiting the generality of the foregoing, each Partner shall execute and deliver any and all inducement letters and other documents required under the Recording Agreement.

5.2 *Time Devoted to Partnership, Outside Activities, and Extended Absences.*

(a) The Partners shall render their services under this agreement to the best of their ability and talents, and shall comply with all reasonable scheduling requirements established by Partnership Vote. The Partners hereby acknowledge and agree that the needs and requirements (including all scheduling requirements) of the Partnership shall take precedence over all other business and personal activities of the Partners.

(b) Notwithstanding anything to the contrary contained herein, the Partners shall be permitted to engage in business activities other than those required hereunder, including, without limitation, (i) songwriting and music publishing activities, as those terms are commonly defined in the music industry; (ii) performing as sidemen musicians or background vocalists for other artists; (iii) producing records of other artists; (iv) managing artists; (v) recording so-called "solo" albums and rendering live performances in connection therewith; and (vi) rendering services of any kind unrelated to the Group or the Group Name in an entertainment medium other than the music field (e.g., acting, directing, writing, commercials, etc.) (hereinafter all such activities shall be collectively referred to as "Outside Activities"); provided, however, that no Partner shall engage in any Outside Activity that would conflict with such Partner's complete performance of all services under this agreement, including, without limitation, the Partnership's complete performance of its obligations under the Recording Agreement.

(c) Except as otherwise provided in this agreement, any and all monies or other assets derived or created by any Partner or Partners during the term hereof as a result of those activities described in paragraph 1.1, but excluding all Outside Activities engaged in by such Partner or Partners in accordance with paragraph 5.2(b) above, shall belong solely to the Partnership; any and all monies or other assets derived or created by any Partner or Partners as a result of Outside Activities engaged in by such Partner or Partners in accordance with paragraph 5.2(b) above shall belong solely to such Partner or Partners.

(d) If a Partner is unable to devote such time and attention to the Partnership business as is necessary to fully and completely perform the services to be performed for the Partnership hereunder due to physical or mental illness or incapacity (unless such illness or incapacity is self-induced), such Partner shall, subject to paragraph 7.2 below, continue to receive its share of the Net Profits of the Partnership for a period of three (3) months from the date such disability commences. If such illness or incapacity continues beyond such three (3) month period, then the remaining Partners may, by Partnership Vote (excluding the disabled Partner), agree upon a full or reduced share of the Partnership's Net Profits and Net Losses for such disabled Partner during the continuance of such illness or incapacity.

5.3 *Music Publishing Participation.*

(a) The Partners hereby acknowledge and agree that each Partner shall be the sole and exclusive owner of an individual publishing company (hereinafter referred to as a Partner's "publishing company designee") which shall solely and exclusively own all of such Partner's copyright interest in and to all musical compositions written or co-written, in whole or in part, by such Partner (hereinafter each such musical composition, or the applicable portion thereof, which is embodied on a recording featuring the performances of the Group [each a "Group Recording"], is referred to as a "Partner Composition").

(b) Each Partner and his respective publishing company designee shall concurrently herewith enter into a Participation Agreement with the Partnership (in the form of Exhibit "B", attached hereto and incorporated herein by this reference) pursuant to which the Partnership shall receive a Percentage (as determined in subparagraph 5.3(c) below) of the so-called "publisher's share" of income derived from exploitation of the applicable Partner's Partner Compositions as embodied on Group Recordings only (hereinafter such income received by the Partnership is referred to as the "Partnership Publishing Participation").

(c) The Percentage, with respect to each particular Group Recording, shall equal ten percent (10%) times the number of Partners in the Partnership at the time such Group Recording was recorded; except the Percentage shall be reduced by ten percent (10%) for any participating Partner who subsequently becomes a Terminated Partner (as hereinafter defined), effective upon the Valuation Date (as hereinafter defined) for such Terminated Partner. For example, if at the time a particular Group Recording was recorded there were four (4) Partners, then the Percentage with respect to the Partner Composition, if any, as embodied on such Group Recording would be forty percent (40%) and if one of such participating Partners later became a Terminated Partner, from and after the Valuation Date for such Terminated Partner the Percentage would be reduced to thirty percent (30%).

(d) Each Partner shall be entitled to a fraction of the Partnership Publishing Participation derived from the Partner Compositions as embodied on Group Recordings recorded while such Partner was a Partner in the Partnership, the numerator of which is "one" and the denominator of which is the then current number of Partners in the Partnership.

(e) For purposes of clarification and notwithstanding anything to the contrary set forth above:

  (i) The Partnership shall have no ownership interest of any kind in the Partner Compositions of the Partners.

  (ii) No Partner shall have any ownership interest of any kind in the Partner Compositions of another Partner.

(iii) The Partnership shall not be entitled to any participation in a Partner Composition under this paragraph 5.3 to the extent attributable to the use of such Partner Composition on a non-Group Recording.

(iv) All obligations of a Terminated Partner to pay to the Partnership such Partner's applicable Partnership Publishing Participation under paragraph 5.3(b) above shall cease as to future earnings from such Terminated Partner's Partner Compositions effective upon the Valuation Date for such Terminated Partner. In addition, such Terminated Partner's right to participate in the Partnership Publishing Participation thereafter received by the Partnership from the other Partner's Partner Compositions shall also cease upon the Valuation Date.

Article VI Management and control

6.1 *Partnership Vote*. Only the Voting Partners (herein defined) shall participate in the control, management, and direction of the Partnership. Except as expressly provided otherwise in this agreement, all matters within the ordinary course of the Partnership business shall be decided by majority vote of the Voting Partners including, in each instance (except as otherwise expressly set forth in this agreement), the vote of any Partner with an interest in the result of such vote (such majority vote is referred to in this agreement as a "Partnership Vote"). All matters outside the ordinary course of the Partnership business shall be decided by, and any acts in contravention of this agreement, may only be authorized by, unanimous vote of the Voting Partners. Until designated otherwise by the unanimous consent of the Voting Partners, the Voting Partners, who shall each have one vote, shall be _____. Upon the Termination (as defined in paragraph 7.3 below) of a Voting Partner, although this paragraph 6.1 shall permit the remaining Voting Partners to elect a replacement Voting Partner by the unanimous consent of the remaining Voting Partners, the remaining Voting Partners shall not be obligated to elect any replacement Voting Partner.

6.2 *Managing Partner*. Pursuant to the unanimous written consent of the Voting Partners, the Voting Partners may, from time to time, designate a managing partner with such powers and duties and subject to removal upon such terms and conditions designated by the Voting Partners. The Partnership's right of consent or approval under any Approved Agreement may be exercised by the managing partner or any other person designated by Partnership Vote.

6.3 *Power to Incur Liabilities*. No Partner shall have authority to bind the Partnership in the ordinary course of its business, by making contracts and/or otherwise incurring obligations in the name or on the credit of the Partnership, except as authorized by Partnership Vote. Any Partner who enters into any contract or incurs any obligation in the name or on the credit of the

Partnership in violation of this paragraph may be held individually liable by the other Partners for the entire amount of the obligation or loss thus incurred by such Partner.

6.4 *Limitations on Authority.* Without the unanimous consent of the Voting Partners, neither the Partnership nor any individual Partner shall have the authority to:

(a) Borrow money in the Partnership's name;

(b) Loan Partnership funds;

(c) Transfer, mortgage, pledge, assign, hypothecate, or otherwise grant, dispose of, or encumber any Partnership assets, including goodwill;

(d) Enter into any agreement that would place a Partner in conflict with, or in breach of, any provision of this agreement, or any Approved Agreement;

(e) Perform any act that would make it impossible to carry on the ordinary business of the Partnership;

(f) In the name of the Partnership, confess a judgment, submit a Partnership claim or liability to arbitration (except as provided in paragraph 6.3 above), endorse any note or act as an accommodation party, or otherwise become surety for any person;

(g) Assign, mortgage, encumber, transfer or sell its share of the Partnership or in the capital assets or property of the Partnership or enter into any agreement as the result of which any person shall become interested with such Partner in the Partnership;

(h) Admit a new partner to the Partnership; or

(i) Bind the Partnership by making contracts or incurring obligations in the name or on the credit of the Partnership, which contracts and/or obligations are outside of the ordinary course of the Partnership business.

6.5 *Personal Liability.* Any Partner who enters into any contract or incurs any obligation on behalf of, or on the credit of, the Partnership in violation of the provisions of this agreement, or otherwise violates the prohibitions of paragraph 6.4 above, shall indemnify the other Partners for any and all losses or expenses incurred pursuant to such contract or obligation.

Article VII **Termination or admission of a partner**

7.1 *Withdrawal of a Partner.*

(a) Any Partner may voluntarily withdraw from the Partnership by giving all the other Partners at least one hundred and twenty (120) days prior written notice of his intention to do so; provided, however, that the Partnership shall have no obligation to continue to use the services of such withdrawing Partner during the one hundred and twenty (120) day period following the date of any such notice. In the event the Partnership elects not to use the services of any withdrawing Partner during all or any remaining portion of the one hundred and twenty (120) day period following the date of notice of any such withdrawal, then the Valuation Date, as hereinafter defined, shall be deemed to be the date on which the Partnership notifies such Partner of its intention not to continue to use such Partner's services.

(b) Should any Partner withdraw from the Partnership as aforesaid and fail or refuse to perform during all or any portion of the required one hundred and twenty (120) day prior notice period (unless such failure or refusal is caused by reasons beyond such withdrawing Partner's control or is expressly requested by the Partnership or agreed to by Partnership Vote, excluding the non-withdrawing Partner), then (i) such Partner shall be liable to the remaining Partners for any and all damages suffered by the Partnership by reason of such failure or refusal (e.g., losses resulting from cancelled concert engagements, television performances, recording sessions, etc.) and (ii) the Valuation Date shall be deemed to be the date on which such failure or refusal commenced.

7.2 *Expulsion of a Partner.* For any reason whatsoever, the Partnership may expel a Partner from the Partnership by unanimous vote of all Voting Partners other than the Partner proposed to be expelled. Such expulsion shall be effective on the day on which the Partnership delivers notice of such expulsion to the expelled Partner.

7.3 *Purchase and Sale of Partnership Interest in the Event of Withdrawal, Expulsion or Death.* Upon the withdrawal, expulsion or death (hereinafter collectively referred to as the "Termination") of any Partner (hereinafter, any such withdrawn, expelled, or deceased Partner shall be referred to as a "Terminated Partner"), such Terminated Partner or such Terminated Partner's estate, as applicable, shall, in accordance with Article VIII, sell to the Partnership, and the Partnership shall purchase, for an amount calculated in accordance with Article VIII, the Terminated Partner's interest in the Partnership; provided, however, that no such obligations shall exist if, upon such Termination, the Partnership elects by Partnership Vote, within one hundred twenty (120) days after such Termination, to liquidate and wind up in accordance with this Agreement.

7.4 *Bankruptcy, Incompetency or Insolvency.* For purposes of this agreement, a judicial declaration of the incompetency, bankruptcy, or insolvency of a Partner shall, absent the unanimous vote of the Voting Partners to the contrary, automatically constitute a withdrawal of such Partner from the Partnership, to which withdrawal the provisions of paragraph 7.1 shall apply, effective as of the last day of the month during which any such event occurs.

7.5 *Continuation of Partnership.* The death, withdrawal, or termination of a Partner shall not cause the winding up of the Partnership as to the other Partners, nor shall it cause any interruption in the conduct of the Partnership business, nor shall it affect the continuity of the Partnership and its business, or the continued use of the Partnership name in the conduct of such business.

7.6 *Restrictions on Transfer.* No Partner shall transfer any or all of such Partner's interest in the Partnership, whether voluntarily or involuntarily (including by way of sale, assignment, transfer to a trust, gift, will, intestate succession, marital dissolution, or death of a spouse), to any person (including any other Partner) without the prior unanimous consent of the Voting Partners. If a Partner purports to transfer any or all of his interest in the Partnership to any transferee without such consent, then such purported transfer shall constitute a withdrawal of such Partner from the Partnership, to which withdrawal the provisions of paragraph 7.1 shall apply, and such purported transferee shall not receive any interest in either the Partnership or any Partnership assets.

7.7 *New Partners.* No new person shall be admitted as a partner to the Partnership except by unanimous vote of all Voting Partners. Any person so admitted to the Partnership as a partner shall (a) agree in writing to be bound by each and every term of this agreement and such additional terms, if any, as shall be determined by the unanimous vote of the Voting Partners, and (b) contribute to the capital of the Partnership such sum, if any, determined by the unanimous vote of the Voting Partners. Unless designated otherwise by unanimous vote of the Voting Partners, any newly admitted Partner shall be deemed a non-voting Partner.

Article VIII Purchase price of a partnership interest

8.1 *Partnership Obligation.* The full purchase price required to be paid by the Partnership for a Terminated Partner's interest in the Partnership shall be an obligation payable solely from Partnership assets (both present and future) and not a personal obligation of the Partners, and shall consist solely of the amounts (regardless of any ongoing royalty participations to which the Partnership may be entitled), and be payable in the manner, specified in this Article VIII.

8.2 *Elimination of Capital Account.*

(a) As of the effective date of the Termination of a Partner (hereinafter the effective date of a Termination [which, in the event of a Partner's death, shall be the date of death] shall be referred to as the "Valuation Date"), the Terminated Partner's individual capital account shall be adjusted to reflect the Terminated Partner's share of Net Profits or Net Losses of the Partnership from the beginning of the fiscal year in which such Termination occurs through the Valuation Date, taking into account for such purpose all Partnership expenses paid or accrued during such period. The Partnership shall use the so-called "interim closing of the books method" for tax and accounting purposes. Such capital account shall be further adjusted by taking into account such Terminated Partner's share of the book value of any tangible assets (or any liabilities) of the Partnership as of the Valuation Date (excluding, however, any interest of the Partnership in any master recordings, videotapes, films or other recordings of any kind embodying the performances of the Group or any one or more of the Partners); provided, however, that in making such computation no value shall be allocated to any interest of such Partner in or to any

Approved Agreement. If, after such adjustment, there remains a credit balance in the Terminated Partner's capital account, then such credit balance shall be paid to the Terminated Partner (if a Terminated Partner is deceased, then any reference in this agreement to either (i) payments to or from, or (ii) statements submitted to, such Terminated Partner shall be interpreted as (I) payments to or from, or (II) statements submitted to, such Terminated Partner's estate and heirs, as applicable) in accordance with paragraphs 8.2(a)(i), (ii), and (iii) below.

(i) Where such credit balance is equal to or less than $ _____, such sum shall be paid, with interest on the unpaid balance at the applicable federal rate as defined in Section 1274 of the Code, prevailing on the Valuation Date (herein referred to as the "Applicable Federal Rate"), as follows: (A) the initial $ _____ _____, plus interest, shall be paid six (6) months from the Valuation Date; and (B) the remaining balance thereof, plus interest, shall be paid twelve (12) months from the Valuation Date. The foregoing payment(s) shall be evidenced by a promissory note executed by the Partnership as thereafter reconstituted.

(ii) Where such credit balance is greater than $ _____ _____, such sum shall be paid, with interest on the unpaid balance at the Applicable Federal Rate as follows: (A) one-half (1/2) or the initial $ _____ _____ thereof, whichever sum is greater, shall be paid in two (2) equal installments, plus interest, six (6) months and twelve (12) months, respectively, from the Valuation Date; and (B) the remaining balance thereof shall be paid in two (2) equal installments, plus interest, eighteen (18) months and twenty-four (24) months, respectively, from the Valuation Date. The foregoing payments shall be evidenced by a promissory note executed by the Partnership as thereafter reconstituted.

(iii) Notwithstanding subparagraphs (a)(i) and (a)(ii) of this paragraph 8.2, the amount of any such promissory notes referenced therein shall be reduced by the amount, if any, of the Terminated Partner's obligations to the Partnership under this agreement, and the Partnership (A) shall always have the option to pay, without penalty, the payments therein provided at a date earlier than required in said subparagraphs, and (B) shall be obligated to use its reasonable efforts to pay each of such payments at the earliest date possible consistent with the best business judgment of the Partnership exercised in good faith and in consultation with the Accountant.

(b) If there remains a debit balance, such debit balance shall be paid to the Partnership by the Terminated Partner at the earlier of (i) the time required pursuant to Treasury Regulation Section 1.704-1(b)(2)(ii)((b)

(3), or (ii) in the same manner as provided in subparagraphs (a)(i) through (a)(iii) of this paragraph 8.2, as if such debit balance were a credit balance. In addition, the Partnership shall have the option, at its sole election, to withhold payment to the Terminated Partner of any sums payable to such Terminated Partner pursuant to paragraph 8.3 below in any amount not in excess of the amount then owing by the Terminated Partner to the Partnership pursuant to this paragraph 8.2. The amount so withheld shall be treated as having been paid to the Terminated Partner pursuant to paragraph 8.3 and then recontributed to the Partnership by the Terminated Partner in satisfaction of the obligation to restore the debit balance in the capital account, and in the case of any obligation of the Terminated Partner to make payments to the Partnership in the future, such withheld amounts shall be treated as made in satisfaction of the obligations to pay amounts due at the time or times furthest in the future.

8.3 *Payment of Future Partnership Income.*

(a) Within forty-five (45) days after receipt by the Partnership of the following revenues only, the Partnership shall pay to the Terminated Partner his applicable pro rata share of such revenues:

(i) *Records.* Royalties and other sums received by, or credited (by reason of a prior advance) to the account of, the Partnership on or after the Valuation Date pursuant to the Recording Agreement on account of the exploitation of master recordings of the Group (whether or not such master recordings embody the performance of the Terminated Partner) which were substantially completed prior to the Valuation Date <u>and</u> while the Terminated Partner was a Partner (hereinafter referred to as "Participation Masters"), following the deduction of (A) any and all cash advances received by the Partnership prior to the Valuation Date <u>and</u> while the Terminated Partner was a Partner, which have not been recouped prior to the Valuation Date, (B) any and all recoupable costs and expenses incurred in connection with the production of such Participation Masters (including recording costs incurred on or after the Valuation Date but in connection with a Participation Master) which have not been recouped prior to the Valuation Date, and (C) any and all other advances and recoupable costs and expenses (e.g. video production costs, "tour support" advances and advances for independent record promotion) received by or incurred on behalf of the Partnership prior to the Valuation Date *and* while the Terminated Partner was a Partner (except that video production costs incurred on or after the Valuation Date but in connection with a Participation Master shall be included in this subsection (C)), which have not been recouped prior to the Valuation Date; provided, however, that (I) the aforesaid royalties payable to any such Terminated Partner shall be computed without regard to any royalty increases or improved royalty calculation provisions in the Recording Agreement secured by the Partnership on or after the Valuation Date, and (II) the aforesaid income of the Partnership and deductions therefrom shall be calculated separately with respect to those Participation Masters recorded under each Recording Agreement and there shall be no cross-collateralization between such Recording Agreements. Whenever Participation Masters are coupled on a record with other master recordings, those royalties attributable to the Participation Masters shall be determined by multiplying the gross royalties derived from such record by a fraction, the numerator of which is the number of Participation Masters on such record and the denominator of which is the total number of master recordings (including the Participation Masters) on such record.

(ii) *Concerts*. Income received by the Partnership after the Valuation Date *and* derived from concert performances in which the Terminated Partner performed less all expenses attributable thereto, including without limitation, costs of travel and accommodations, equipment, rental lighting and sound, opening acts, stage crew, musicians, singers and other personnel, publicity and promotion;

(iii) *Merchandise*. Income received by, or credited (by reason of a prior advance) to the account of, the Partnership after the Valuation Date and derived from any tour and/or retail merchandising agreements entered into by the Partnership prior to the Valuation Date and while the Terminated Partner was a Partner, but solely with respect to sales occurring thereunder prior to the Valuation Date, following the deduction of (A) all recoupable costs and expenses paid or incurred under such agreements prior to the Valuation Date and while the Terminated Partner was a Partner, which have not been recouped prior to the Valuation Date, and (B) all cash advances received by the Partnership under such agreements prior to the Valuation Date and while the Terminated Partner was a Partner, which have not been recouped prior to the Valuation Date; and

(iv) Partnership Publishing Participation. Partnership Publishing Participation received by the Partnership on or after the Valuation Date and derived from Partner Compositions as embodied on Group Recordings recorded before the Valuation Date and while the Terminated Partner was a Partner, following the deduction (in the following order and to the extent not already deducted hereunder) of (A) any and all cash advances received by the Partnership in connection with the applicable Partner Compositions prior to the Valuation Date and while the Terminated Partner was a Partner which have not been recouped prior to the Valuation Date, and (B) any and all recoupable costs and expenses incurred with respect to the applicable Partner Compositions which have not been recouped prior to the Valuation Date.

(v) *Miscellaneous*. Income received by, or credited (by reason of a prior advance) to the account of, the Partnership after the Valuation Date and derived from any materials, projects or items other than those described in paragraphs 8.3(a)(i), (ii) or (iii) above (e.g., motion pictures, videos, television and radio programs and master recordings recorded for a motion picture soundtrack album) substantially created or completed by the Partnership prior to the Valuation Date *and* containing the Terminated Partner's performances and/or creative input as a member of the Group, following the deduction of (A) all recoupable or deductible production costs and expenses (including all expenses of exploitation) paid or incurred in connection with such materials, projects or items, which have not been recouped or deducted prior to the Valuation Date, and (B) all cash advances received by the Partnership in connection with such materials, projects or items prior to the Valuation Date <u>and</u> while the Terminated Partner was a Partner, which have not been recouped prior to the Valuation Date. If any such materials, projects or items are coupled with other materials, projects or items in which the Terminated Partner does not participate hereunder and are exploited together as a package (e.g. a compilation video), a fair allocation of the income derived therefrom shall be made in good faith by the Accountant.

(b) For purposes of the foregoing paragraph 8.3(a): (i) Each agreement, material, project and/or item under paragraphs 8.3(a)(iii), (iv), and (v) above shall be treated separately and shall not be cross-collateralized in calculating what is due the Terminated Partner; (ii) prior to the calculation of the Terminated Partner's share of each respective source of income described in paragraphs 8.3(a)(i), (ii), (iii), (iv), and (v) above, there shall first be deducted from the otherwise applicable amount of such income in question any and all collection costs (including attorneys' fees), accounting and audit costs, bona fide third party payments (including without limitation, record producer's, recording engineer's and video director's royalties), production, exploitation and other costs expended or incurred by or on behalf of the Partnership and/or the Partners (or any of them), legal fees and costs, and management, business management and agency commissions, which are attributable thereto (collectively "Expenses"); and (iii) the Terminated Partner's "pro rata share" shall equal a fraction, the numerator of which is "one" and the denominator of which is the total number of Partners participating in the particular source of income in question.

(c) As provided in paragraph 5.3 above, and notwithstanding anything else to the contrary set forth above, a Terminated Partner shall not participate in any Partner Publishing Income received by the Partnership more than six (6) months after the Valuation Date.

8.4 *Audits*. In the event the Partnership shall undertake an audit or examination of the books and records of any payor of royalties or any other sums referred to in paragraphs 8.3(a)(i), (ii), (iii) and (iv) above, the Terminated Partner shall receive the Terminated Partner's applicable share hereunder of any resulting recovery within forty-five (45) days after the Partnership's receipt thereof, after first subtracting therefrom that portion of all applicable Expenses attributable thereto determined by multiplying the amount of such Expenses by a fraction, the numerator of which is the Terminated Partner's share of such recovery and the denominator of which is the total amount of such recovery.

8.5 *Overpayments.* Notwithstanding anything to the contrary contained herein, if, at any time after the Valuation Date, the actual income of the Partnership (as reflected on statements received by the Partnership, audits conducted on behalf of the Partnership, or otherwise) indicates that any Terminated Partner received, prior to the Valuation Date, an allocation of an advance or other income described in Article IV that is disproportionately greater than such Terminated Partner's share of such actual income, then, to the extent the Partnership is required to repay all or a portion of the applicable advance or other income, such Terminated Partner shall promptly repay to the Partnership the amount of the excess advance or other income paid to such Terminated Partner. In the event that a Terminated Partner fails to make any payment required by this paragraph 8.5, then, without limiting any of the Partnership's other remedies, the Partnership shall have the right to deduct the amount of such payment from any and all monies otherwise payable to the Terminated Partner.

8.6 *Goodwill.* Upon any Termination or dissolution of the Partnership, the goodwill, if any, of the Partnership, including any goodwill associated with the Group Name, shall <u>not</u> be considered an asset of the Partnership for purposes of valuation or division of the Partnership assets under this Article VIII or under Article X below.

8.7 *Complete Termination of Partner's Interest.* The payment of the amounts set forth in this Article VIII shall constitute a complete buyout, settlement, and liquidation of any and all right, title, and interest that a Terminated Partner may have in, to, or against the Partnership or the Partners. Without limiting the generality of the foregoing, except as specifically provided in this Article VIII, no Terminated Partner shall have any interest in income received by the Partnership after the Valuation Date. Payments under paragraph 8.2 shall be treated as payments for a Terminated Partner's interest in Partnership property, as described in Section 736(b) of the Code. Payments under paragraph 8.3 shall be treated as payments of the Terminated Partner's distributive share of Partnership income, as described in Section 736(a) of the Code.

Article IX **Additional provisions regarding termination**

9.1 *Statements and Inspection of Records.* In connection with ongoing payments that may be required to be made to a Terminated Partner pursuant to Article VIII, the Accountant shall furnish to the Terminated Partner semi-annual statements regarding any such payments made during the prior six (6) month period, and the basis of such payments. Each such statement shall be binding upon the Terminated Partner, unless (a) an objection is made in writing stating the basis therefor and notice of such objection is delivered to the Partnership within three (3) years after the date of such statement, and (b) if the Partnership denies the validity of such objection, then suit is instituted within one (1) year after the date notice of such objection is delivered to the Partnership. During said three (3) year period, upon reasonable advance notice and at reasonable times, the Terminated Partner, shall, at the Terminated Partner's sole cost and expense, have the right to inspect the portion of the Partnership's books and records that relate to monies payable to such Terminated Partner.

9.2 *Third Party Books and Records.* It is agreed and understood that in rendering the statements and making the payments required by Articles VIII and IX hereof, the Partnership shall rely on statements it receives from third parties; accordingly, the Partnership shall have no liability to any Terminated Partner if such payments or statements are accurately based on the information supplied to the Partnership by such third parties. The Partnership shall make available to the Terminated Partner copies of the portions of such third party statements that relate to monies payable under Articles VIII and IX hereof. Moreover, if the Partnership does not, prior to the Trigger Date (hereinafter defined), conduct an examination or audit of any particular third party's books and records in respect of a particular royalty or income statement rendered by such third party, then the Terminated Partner shall have the right, by written notice to the Partnership during the four (4) month period following the Trigger Date, to require the Partnership to conduct, at the Terminated Partner's sole expense and with a certified public accountant of the Terminated Partner's choice, such an audit or examination (hereinafter referred to as a "Termination Audit"). The Partnership shall, within thirty business (30) days after the Partnership receives any monies from any third party pursuant to a particular Termination Audit, pay to the Terminated Partner an amount equal to the "Termination Share." As used herein, the term "Termination Share" shall mean the amount of Net Monies (hereinafter defined) multiplied by a fraction, the numerator of which is the amount of money paid by such third party to the Partnership as a result of such Termination Audit that is allocable to the Terminated Partner pursuant to this Agreement, and the denominator of which is the total amount of money paid by such third party to the Partnership as a result of such Termination Audit. As used herein, the term "Net Monies" shall mean the gross monies paid by third parties to the Partnership pursuant to a Termination Audit, less any and all expenses incurred by the Partnership or the Terminated Partner pursuant to the foregoing. The party incurring any such expenses shall be reimbursed therefor from the first Net Monies received by the Partnership. As used herein, the term "Trigger Date" shall mean that date four (4) months before the last date on which the Partnership shall, under the Partnership's agreement with any such third party, have the right to conduct an examination of the applicable books and records of such third party.

9.3 *Damages and Liabilities.* Notwithstanding anything to the contrary contained in this agreement, (a) in the event a Partner is expelled by the Partnership for a breach of a material provision of this agreement, excluding the withdrawal of a Partner earlier than permitted under paragraph 7.1(a) (which is subject to the provisions of paragraph 7.1(b)), then such Partner (or his estate and heirs, as applicable) shall be liable for any damages suffered by the Partnership from such material breach, and (b) each Terminated Partner (or his estate and heirs, as applicable) shall be liable for payment to the Partnership for such Partner's allocable share of any claim, action, liability, damage, cost, or expense arising out of, or in connection with, any Participation Masters, or any materials, projects, or transactions acquired, produced, consummated, or completed by the Partnership on or before the Valuation Date ("Pre-Valuation Liabilities"). Such Partner's allocable share of Pre-Valuation Liabilities shall be based on such Partner's allocable share of Net Profits and Net Losses at the time such Pre-Valuation Liabilities were incurred. Without limiting any of the Partnership's other rights or remedies, any such damages or Pre-Valuation Liabilities shall be (a) deducted from the amount of any monies that the Partnership is required to pay to the Terminated Partner under this agreement, and/or (b) added to the amount of any monies that the Terminated Partner is required to pay to the Partnership under this agreement.

9.4 *Assignment of Former Partner's Rights.* Except as expressly provided in Article VIII, a Terminated Partner (or his estate and heirs, as applicable) shall not have any rights, claims, or interest in or to (a) any master recordings recorded either before or after the Valuation Date under or pursuant to the Recording Agreement or otherwise; (b) any other tangible or intangible assets of the Partnership, including the Group Name and any goodwill, whether created or acquired before or after the Valuation Date; or (c) any proceeds derived from or arising out of the foregoing. Accordingly, and without limiting the generality of the foregoing, effective immediately upon the Valuation Date, the Partnership shall become the assignee of all such Terminated Partner's right, title, and interest in and to all tangible and intangible assets of the Partnership, including all master recordings, the Group Name, any logos utilized by the Group, the Recording Agreement and all other agreements to which the Partnership is a party, and any goodwill, and all proceeds thereof. Further, the Terminated Partner (or his estate and heirs, as applicable), shall have no rights, claims or interest whatsoever in and to any future assets of the Partnership created or acquired after the Valuation Date, including without limitation, any agreements to which the Partnership may thereafter become a party or any master recordings (audio and/or visual), in any medium, of the Group (as thereafter reconstituted) or any member thereof, together with all proceeds therefrom, recorded after the Valuation Date.

9.5 *The Group Name.*

(a) From and after the Valuation Date, neither the Terminated Partner nor, if applicable, representatives or heirs thereof shall avail themselves of or use the Group Name or any substantially similar name or designation in any medium or commercial manner whatsoever; provided, however, that, notwithstanding anything to the contrary in this agreement, any Terminated Partner shall have the right to refer to himself as being formerly a member of "_____"; provided, however, that no Terminated Partner shall refer to himself as being formerly a member of "_____" in any manner that would confuse or mislead a reasonable person into believing that such Terminated Partner is a member of "_____" at the time of, or at any time after, such reference. From and after the Valuation Date, the Partnership, as thereafter reconstituted, shall be the sole and exclusive owner of all right, title and interest, including without limitation, the trademark and service mark, in and to the Group Name and shall have the continuing and unrestricted right in and to the exclusive use of the Group Name and all substantially similar designations in any medium or commercial manner whatsoever.

(b) Notwithstanding anything to the contrary in this agreement, if the Partnership dissolves and winds up or otherwise ceases to operate, then all rights in and to the Group Name, including the right to use the Group Name as a professional recording and/or performing group, shall vest jointly in the Voting Partners who are Partners at the time the Partnership ceases to operate. The Group Name shall thereafter be administered and controlled by majority vote of such Voting Partners who are then alive. If such Voting Partners are unable to decide by majority vote regarding a particular use of the Group Name, then no such Voting Partner shall have the right to such particular use of the Group Name.

9.6 *Name and Likeness of Former Partner.*

(a) From and after the Valuation Date, the Partnership shall have the continuing and unrestricted nonexclusive right to use the name, photograph, likeness, voice and biographical materials of the Terminated Partner on or in connection with, *inter alia*, the following:

(i) All records and tapes manufactured from Participation Masters and any other master recordings (audio and/or visual) embodying, in whole or in part, the performances of the Terminated Partner;

(ii) All musical compositions written in whole or in part by the Terminated Partner and recorded or partially recorded (whether or not released) by the Partnership; and

(iii) All exploitations of any projects or items, including without limitation, merchandising items, television programs, motion pictures and videos, in which the Terminated Partner participated.

(b) In no event shall the Termination of a Partner in any way affect or detract from any grant made by the Partnership and/or the Terminated Partner before the Valuation Date of rights in and to the results and proceeds of the services of such Terminated Partner, or the rights to use such Terminated Partner's name, likeness, photograph, voice, or biographical materials.

9.7 *Assumption of Partnership Obligations.* Upon any purchase of a Partnership interest pursuant to this agreement, the Partnership shall assume all Partnership obligations incurred after the Valuation Date, and shall protect and indemnify the Terminated Partner (or his estate and heirs, as applicable) from liability for any such obligations. The preceding sentence shall not relieve a Terminated Partner (or his estate and heirs, as applicable) of such Terminated Partner's share of Pre-Valuation Liabilities. Accordingly, a Terminated Partner (or his estate and heirs, as applicable) shall remain liable for such Terminated Partner's share of Pre-Valuation Liabilities.

9.8 *Publication of Notice.* Upon the purchase of a Partnership interest pursuant to this agreement, the Partnership shall, at its own expense, and as soon as reasonably practicable, cause to be prepared, published, filed, and served all such notices as may be required by law to protect a Terminated Partner (or his estate and heirs, as applicable) against liabilities incurred by the Partnership after the Valuation Date.

9.9 *Nondisclosure.* The Partners specifically agree that neither they nor their agents, representatives, estate or heirs shall at any time after the execution hereof knowingly provide to any third party any information concerning this agreement, other than as shall be necessary in connection with the exercise of rights granted herein or as required by law. Promptly following the withdrawal or expulsion of a Terminated Partner, the Partnership and such Terminated Partner (or his estate and heirs, as applicable) shall issue a joint press release regarding such withdrawal or expulsion, which press release shall be approved by Partnership Vote and by the Terminated Partner (or said estate and heirs). If, however, the parties shall be unable to agree upon the form and content of such a press release, then no press release shall be issued. Neither the Partnership nor any Partner shall disseminate any information or make any statements that would reflect in any negative manner on the Terminated Partner (or his estate and heirs, as applicable) and the Terminated Partner (or his estate and heirs, as applicable) shall not disseminate any information or make any statements that would reflect in any negative manner on the Partnership or any Partners. Furthermore, the Terminated Partner shall not, without the consent of the Partnership, give any interviews or write, prepare or assist in the preparation of any books, articles or other writings disclosing confidential information in respect of the Partnership, the Partners or any activities thereof.

Article X **Dissolution**

10.1 *Dissolution in Accordance with Agreement.* The Partnership shall be dissolved in accordance with this agreement upon the occurrence of any of the following events:

(a) The mutual agreement of all Voting Partners;

(b) The death of any Partner;

(c) The withdrawal of any Partner if such withdrawal is not in violation of the terms of this agreement;

(d) The expulsion of any Partner if such expulsion is not caused by the expelled Partner's breach of a material provision of this agreement; or

(e) The Partnership's inability to resolve a deadlocked issue if such inability prevents the Partnership from conducting business for sixty (60) consecutive days.

10.2 *Dissolution Not in Accordance with Agreement.*

(a) No Partner shall commit any "Wrongful Act," which term shall be defined as one or more of (i) a breach by a Partner of any material provision of this agreement, (ii) conduct by a Partner that results in a dissolution not specified in paragraph 10.1, or (iii) conduct by a Partner that permits a court to dissolve the Partnership pursuant to Tennessee Code Annotated Section 61-1-801. Any Wrongful Act committed by a Partner shall give the other Voting Partners, for a reasonable period of time following such breach, the right to dissolve the Partnership by Partnership Vote. The foregoing shall not preclude a court of competent jurisdiction from dissolving the Partnership pursuant to Tennessee Code Annotated Section 61-1-801.

(b) If the dissolution of the Partnership shall result from a Wrongful Act, then the provisions of the Tennessee Revised Uniform Partnership Act (Tennessee Code Annotated Sections 61-1-101 *et seq.*, including Tennessee Code Sections 61-1-703 and 61-1-806 and any other provisions thereof regarding wrongful dissolutions) shall apply to any such dissolution, and the Partner committing such Wrongful Act (the "Breaching Partner") shall be deemed to have wrongfully caused the dissolution of the Partnership. Notwithstanding the foregoing, if, following any dissolution resulting from a Wrongful Act, the Partnership liquidates and winds up, then the provisions of paragraph 10.4 shall apply to such liquidation and winding up, subject to the right of the Partners who are not Breaching Partners to recover from the Breaching Partner damages for breach of this agreement, whether pursuant to Section 61-1-806 of the Tennessee Code or otherwise.

10.3 *Continuation of Business.* Notwithstanding any other terms of this agreement, upon the dissolution of the Partnership other than by reason of those events specified in paragraphs 10.1(a) and (e), then the Partners who are not Terminated Partners shall have the right to continue the Partnership if such continuation is approved by Partnership Vote.

10.4 *Liquidation.*

(a) If the Partnership is dissolved and the Partnership business is not continued pursuant to paragraph 10.3, then the remaining Partners shall proceed with an orderly liquidation, distribute the assets of the

Partnership as provided in this paragraph 10.4, and otherwise wind up the affairs of the Partnership. Prior to any distribution of assets to the Partners, the individual capital accounts of the Partners shall be adjusted to the date of dissolution to reflect net profit and net loss accrued or incurred from the date of the last accounting to the date of dissolution. Following such adjustments, except as provided in paragraph 9.5 with respect to the Group Name and paragraph 10.4(b) with respect to intangible assets of the Partnership, the assets of the Partnership shall be liquidated and the proceeds from liquidation shall be distributed as follows: (a) first, to creditors other than the Partners in the order of priority as provided by law, (b) second, to pay the outstanding balance of any loans or advances made to the Partnership (including accrued interest) by the Partners, and (c) third, to the Partners in proportion to their positive capital account balances. If upon liquidation, a Partner's capital account has a debit balance, then such Partner shall restore to the Partnership the amount of such debit balance in accordance with Treasury Regulation Section 1.704-1(b)(2)(ii)(b)(3), namely, by the later of (i) the end of the taxable year of liquidation, or (ii) 90 days from the date of liquidation.

(b) Notwithstanding the provisions of paragraph 10.4(a), if the Partnership is dissolved and not continued pursuant to paragraph 10.3, then the Partnership's intangible assets and any monies derived therefrom shall be controlled and administered by the then living Voting Partners existing as of the date of the event that resulted in the dissolution and winding up of the Partnership. Such administration shall be conducted by a majority vote of such then living administrating Partners, and shall be subject to any agreements entered into by the Partnership before the date of dissolution of the Partnership.

Article XI **Miscellaneous**

11.1 *Attorneys' Fees.* If any action, suit, or other proceeding, controversy, or dispute arises that is based on or related to this agreement, then the party prevailing therein shall be entitled to recover, as an element of its cost of suit, separate and apart from damages, all reasonable attorneys' fees and court costs incurred therein, whether or not such action, suit, or proceeding proceeds to final judgment. No such sum for attorneys' fees and/or court costs shall reduce the amount of any judgment, and the amount of any judgment shall not be considered in determining whether such fees and/or court costs are reasonable.

11.2 *Notices.* All notices to be given pursuant to this agreement shall be in writing and shall be delivered by hand or sent by United States certified mail, postage prepaid, return receipt requested, or sent by facsimile machine with a copy contemporaneously sent by United States certified mail, postage pre-paid, return receipt requested; provided, however, that any statements required under paragraph 9.1 may be sent by regular mail. Properly addressed notices delivered or sent as provided herein shall be deemed given when delivered by hand, or when postmarked if delivered by mail, or on the date thereof if sent by facsimile machine. Any such notice shall be deemed properly addressed if sent to the last known address of the Partnership or of the Partner to whom such notice is to be given or to the address designated by either such party in writing. Until otherwise designated in writing, any such notices shall be sent to the parties at the following addresses:

_____   _____
_____   _____
_____   _____
_____   _____
_____   _____
_____   _____

Copies   of   all   such   notices   to   the   Partnership   shall   be   sent   to
_____.

11.3 *Additional Documents and/or Acts.* The Partners hereby agree that they will execute any further documents and perform any acts that are now or may become necessary to effectuate the terms of this agreement.

11.4 *Amendment of Agreement.* This agreement may be amended only by written instrument signed by all of the Partners.

11.5 *Waiver; Cumulative Remedies.* No waiver by any party, whether express or implied, of any provision of this agreement or any default hereunder shall affect such party's right to thereafter enforce such provision or to exercise any right or remedy in the event of any other default. All rights and remedies at law or equity, or pursuant to any provision of this agreement, that any party may enjoy as a result of the default or breach of this agreement by any other party, shall be deemed cumulative.

11.6 *Interpretation.* Unless otherwise specified herein or unless the context otherwise clearly requires, (a) the masculine gender used herein includes the feminine and neuter genders; (b) the grammatically plural form of any term defined in the singular form hereunder shall also be the plural form of such defined term, and the grammatically singular form of any term defined in the plural form hereunder shall also be the singular form of such defined term; and (c) the word "including" shall mean "including, but not limited to." The headings of the paragraphs of this agreement are for convenience of reference only and shall not be deemed to limit or in any way affect the scope, meaning, or intent of this agreement or any part hereof. The word "person" shall refer to any individual, corporation, partnership, association, trust, or other entity, or any other organized group of individuals or legal successors or representatives of the foregoing.

11.7 *Construction and Validity*. This agreement shall be construed in accordance with the laws of the State of Tennessee applicable to agreements performed wholly within such State. The illegality or unenforceability of any term of this agreement shall not affect the validity of the remaining portion of this agreement.

11.8 *Entire Agreement*. This agreement is intended by the parties hereto as the final expression of their agreement and understanding with respect to the subject matter hereof and as a complete and exclusive statement of the terms thereof; this agreement supersedes any and all prior or contemporaneous agreements and understandings related thereto including the Prior Agreement.

11.9 *Binding Upon Heirs, Successors, Representatives, and Assigns*. This agreement shall be binding upon, and shall inure to the benefit of, each of the parties hereto and their respective heirs, successors, representatives, and permitted assigns.

11.10 *Preparation of this Agreement*. THIS AGREEMENT WAS PREPARED ON BEHALF OF THE PARTNERSHIP BY _____ _____, AND ALL PARTIES HERETO HAVE HERETOFORE VOLUNTARILY CONSENTED TO THE PREPARATION OF THIS AGREEMENT ON BEHALF OF THE PARTNERSHIP. EACH PARTY HERETO HAS BEEN ADVISED AND UNDERSTANDS THAT CERTAIN INHERENT CONFLICTS EXIST BETWEEN THE PARTIES HERETO AS EACH PARTY MAY HAVE DIFFERENT NEEDS OR DESIRES IN THE RELATIONSHIP BEING STRUCTURED IN THIS AGREEMENT WHICH COULD BEST BE REPRESENTED BY AN ADVISOR REPRESENTING SUCH PARTY's INTERESTS ONLY. IN THIS REGARD, EACH PARTY UNDERSTANDS THAT EACH SUCH PARTY HAS THE RIGHT TO BE REPRESENTED BY SEPARATE AND INDEPENDENT COUNSEL IN CONNECTION WITH THIS AGREEMENT AND LOEB HAS ADVISED EACH SUCH PARTY TO SECURE SUCH SEPARATE REPRESENTATION. EACH SUCH PARTY HAS HAD FULL AND AMPLE OPPORTUNITY TO SECURE SUCH SEPARATE AND INDEPENDENT REPRESENTATION BUT HAS CHOSEN NOT TO DO SO.

IN WITNESS WHEREOF, the Partners have executed this General Partnership Agreement as of the date first written above.

_____   _____
[NAME]   [NAME]

_____   _____
[NAME]   [NAME]

Exhibit "A" <u>**Assets owned by prior partnership**</u>

Recording contract

This form is provided only as a guide to students of artist management for the recording industry. Though it gives a general view of components of a contract between an artist and a record company, the preparation and execution of an actual agreement requires the guidance of separate attorneys to assure that the final version is representative of the agreements and commitments made by the artist and the record company.

Recording agreement

THIS AGREEMENT is made as of this _____ day of _____, _____, between _____ **d/b/a** _____ **Records** (hereinafter referred to as "Company"), _____, and _____ (hereinafter referred to as "you"), _____.

1. *Term*

    1.01. The term of this agreement and the initial Contract Period hereunder will begin on the date hereof. Each Contract Period hereunder will end, unless extended as provided herein, on the *later* of (a) the last day of the twelfth (12th) month following the month of your Delivery to Company or (b) the last day of the ninth (9th) month following the month of Company's United States initial retail street date, of the last Album Delivered by you in fulfillment of your Recording Commitment for the Contract Period concerned.

    1.02. (a) You hereby grant Company six (6) options to extend the term of this agreement for one additional Contract Period each (each an "Option Period"), on the same terms and conditions applicable to the initial Contract Period except as otherwise provided herein. Company may exercise each such option by giving you notice at any time before the expiration of the Contract Period then in effect. If Company exercises an option, the applicable Option Period will begin immediately after the end of the then current Contract Period (or, if Company so advises you in its exercise notice, such Contract Period will begin on the date of such exercise notice).

        (b) Notwithstanding anything to the contrary contained in this paragraph 1.02, if Company has not exercised its option to extend the term of this agreement for an additional Contract Period as of the date the then-current Contract Period would otherwise expire, the following will apply:

   (i) You will notify Company (the "Option Warning") that the applicable option has not yet been exercised.

   (ii) Company will have the right to exercise such option at any time until the date ten (10) business days after its receipt of the Option Warning (the "Extension Period").

   (iii) The then-current Contract Period will continue in effect until either the end of the Extension Period, or Company's notice to you ("Termination Notice") that Company does not wish to exercise such option, whichever is sooner.

   (iv) For the avoidance of doubt, nothing herein will limit Company's right to send a Termination Notice to you at any time, nor limit Company's right to exercise an option at any time if you fail to send Company an Option Warning in accordance with clause 1.02(b)(i) above.

2. *Delivery Obligations*

2.01. During each Contract Period you will Deliver to Company commercially and technically satisfactory Masters. Such Masters will embody your featured vocal performances of contemporary selections, not recorded "live" or "in concert," and that have not been previously recorded by you, whether hereunder or otherwise. (Any Masters that were partially or completely recorded prior to the term of this agreement, and which are delivered by you to Company will be deemed to have been recorded during the initial Contract Period.) Neither Multiple Record Albums nor Joint Recordings may be recorded as part of your Recording Commitment hereunder without both your and Company's written consent. Without limiting the foregoing, Company has the right to reject any Master that Company reasonably believes is either offensive to reasonable standards of public taste or in violation of the rights of others.

2.02. During each Contract Period, you will perform for the recording of Masters and you will Deliver to Company those Masters (the "Recording Commitment") necessary to satisfy the following schedule (sometimes respectively, the "First Album," "Second Album," "Third Album," "Fourth Album," "Fifth Album" "Sixth Album" and "Seventh Album"):

Contract Period	Recording Commitment
Initial Contract Period	one (1) Album
Option Periods	one (1) Album each

2.03. The Album in fulfillment of your Recording Commitment for the initial Contract Period will be Delivered to Company on or before _____, _____, and the Album in fulfillment of your Recording Commitment for each Option Period will be Delivered to Company within four (4) months after the commencement date of the Option Period concerned.

2.04. You will not deviate from the Delivery schedule specified in paragraph 2.03 without Company's written consent; timely Delivery as provided therein is a material obligation hereunder. You agree not to commence the recording of any Record hereunder until nine (9) months after the date of Delivery to Company of the immediately preceding Record in fulfillment of your Recording Commitment hereunder. Each Record will consist entirely of Masters made in the course of that recording project.

2.05. (a) Unless Company requests that you to deliver the Masters in additional and/or other form(s), which Company may do in its sole discretion, and you hereby agree to comply with any such requests:

   (i) you agree to Deliver to Company each Master hereunder in the form of a Digital Master. You will concurrently deliver all two inch and half inch tapes or Pro Tools Files and all multitrack tapes recorded in connection with the recording project, including, without limitation, all twenty-four (24) track master tapes;

   (ii) upon Company's request, you agree to Deliver a 96Khz/24 bit 2 channel stereo version and a 5.1 channel surround sound version of each Recording embodied on a Master hereunder for use on DVD Audio discs (which such versions shall be created pursuant to a recording budget separate from the Authorized Budget applicable to such Master and shall not be payable from or reduce any Authorized Budget); and

   (iii) you agree to Deliver to Company any additional materials as may be required pursuant to the master delivery requirements of Company in effect at such time.

(b) You shall comply with Company's policies with respect to samples, and you hereby warrant and represent that all information supplied by you to Company in that regard is and shall be complete and correct. As of the date hereof, Company's policies with respect to all samples embodied in any Master Recording (including remixes of Master Recordings, regardless of whether such remixes will be commercially released) are as follows:

   (i) Prior to Company's authorization of pre-mastering (e.g., equalization and the making of reference dubs or the equivalent thereof in the applicable configurations) for a particular set of Master Recordings hereunder, you shall deliver the following to Company for the applicable set of Master Recordings:

   (A) A detailed list of any and all samples embodied in each Master Recording.

   (B) A written clearance or license for the perpetual, non-restrictive use of each such sample interpolated in each Master

Recording in any and all media from the copyright holder(s) of the Master Recording and the Composition sampled.

(C) Any and all necessary information pertaining to credit copy required by the copyright holder(s) of each sample interpolated in each Master Recording.

(ii) No Master Recording will be scheduled for release and no Master Recording shall be deemed to be Delivered to Company hereunder (and no Advances due on Delivery, if any, will be paid) until such written sample clearances (including credit copy, if any) have been obtained and approved by Company.

(iii) If any such sample clearance provides for an advance, a flat-fee "rollover" payment and/or a royalty payment for Net Sales of the applicable Master Recording and your record royalty account hereunder is in an unrecouped position at the time such royalties are due, then, notwithstanding anything to the contrary contained herein, you shall be solely responsible for making, and shall make, such payment(s) to the applicable Person promptly upon receipt from Company of such Person's accounting statement thereof. If Company makes any such payment(s), such payment(s) will constitute an Advance and will be recoupable from all monies becoming payable by Company to you or the Artist under this agreement.

(c) Provided you have complied with your other material obligations hereunder and Company is in receipt of all items described in paragraph 14.13 below, the date of Delivery of a Record in fulfillment of your Recording Commitment will be the date of receipt of such Digital Master by Company at the address specified on page 1 hereof; concurrently therewith, you will send a written notice to Company that you have so delivered, however, failure to send such notice shall not be deemed to be a breach of this agreement.

2.06. Company's election to make a payment to you which was to have been made upon Delivery of Masters or to release a Record derived from such Masters will not be deemed to be its acknowledgment that such "Delivery" was properly made, and Company will not be deemed to have waived either its right to require such complete and proper performance thereafter or its remedies for your failure to perform in accordance therewith. If Company chooses to release any Album in fulfillment of your Recording Commitment hereunder prior to the completion of all requirements of Delivery of such Album, Company will pay any Advance due upon the Delivery of such Album promptly after the United States release of such Album.

2.07. Company may release up to three (3) "Greatest Hits," "Best of," or other similar compilation albums during the term, subject only to consulting with you as to the repertoire to be included. If Company so requests, you shall deliver to Company within ninety (90) days following such request, up to three (3) Masters, consisting of newly recorded material, for inclusion on any such albums. Such Masters will be recorded in accordance with the procedures set forth in paragraph 3 below and all costs thereof shall constitute Advances. No "Greatest Hits," "Best of," or other similar compilation shall be deemed part of your Recording Commitment hereunder.

3. *Recording Procedure*

3.01. You will conduct recording sessions only after first obtaining Company's written approval of the individual producer, the places of recording, the Compositions to be recorded and the Authorized Budget (defined below). You will request such approvals at least fourteen (14) days prior to the proposed first date of recording, and Company will not unreasonably withhold any such approval. If Company disapproves any of the foregoing, you will promptly submit alternative proposals, but in all instances you will allow Company a reasonable period of review prior to the proposed first date of recording.

3.02. You will engage all artists, producers, musicians, and other personnel for the recording sessions hereunder, but only after Company sets the budget for all Recording Costs to be incurred in connection therewith, to be determined by the Company, in its sole discretion, after meaningful consultation with you (the "Authorized Budget"). The Authorized Budget may provide for payment to you of no more than union scale for your services (subject to paragraph 4.01) and will not contain a charge for arrangements or orchestrations supplied by you. The Authorized Budget will constitute the maximum amount that you may expend for the applicable session or sessions. The granting of authorizations and the approval of Authorized Budgets are entirely within Company's discretion. Company has the right to have a representative attend all recording sessions conducted pursuant to this agreement at Company's cost on a non-recoupable basis. Without limiting Company's other rights or remedies, if it reasonably appears to Company that the unpaid Recording Costs for any Masters will exceed the amount remaining in the Authorized Budget, Company has the right to immediately cease paying sums from the Authorized Budget unless you establish to Company's reasonable satisfaction that you can and will pay or reimburse Company for any Recording Costs in excess of the Authorized Budget. Nothing contained in this agreement will be deemed to make you Company's agent or authorize you to incur any costs on Company's behalf under this agreement. Notwithstanding the foregoing, without your consent, the minimum Authorized Budgets for Records recorded pursuant to the Recording Commitment will not be less than the following:

Album	Minimum Budget
First	$ _____
Second	$ _____
Third	$ _____
Fourth	$ _____
Fifth	$ _____
Sixth	$ _____
Seventh	$ _____

3.03. You will deliver to Company copies of substantiating invoices, receipts, Form Bs, vouchers and similar satisfactory documentary evidence of Recording Costs for the production of each particular Record of the Recording Commitment, and if you fail to do so, Company's obligation to pay further sums from the Authorized Budget will be suspended until delivery thereof. You agree to deliver to Company (or cause the individual producer of the Masters to deliver) the Immigration and Naturalization Service certificates described in paragraph 3.05 below. Form Bs and W-4s to Company within seventy-two (72) hours after each session hereunder and you shall timely make all required union payments. You further agree to deliver all other invoices, receipts, vouchers and documents within one (1) week after your or the producer's receipt thereof. If Company pays any late-payment penalties solely by reason of your failure to make timely delivery of any such materials, you will reimburse Company for same upon demand and, without limiting its other rights and remedies, Company may deduct an amount equal to all such penalties from monies (other than Mechanical Royalties) otherwise payable to you under this agreement. Company will be responsible for late-payment penalties only if caused solely by Company's acts or omissions. You agree, represent and warrant that all Masters delivered by you to Company hereunder will be free and clear of any liens, encumbrances or claims by any Person (other than those, if any, resulting from Company's acts or omissions), and all Recording Costs with respect thereto shall have been fully paid (other than those unpaid as a result of the acts or omissions of Company).

3.04. (a) Without limiting the foregoing, your obligations include furnishing the services of the individual producers of Masters hereunder, and you are responsible for engaging and paying for such services.

(b) In the event that Company agrees that Company shall engage and/or directly pay the producer of any Master on your behalf (provided that Company is under no obligation whatsoever to agree), the following will apply:

(i) Your royalty account and the Authorized Budget for the recording project concerned will be charged with a Recording Cost item in the amount that Company is obligated to pay such producer in connection with that project.

(ii) The royalties (excluding Mechanical Royalties) payable to you in respect of those Recordings pursuant to Article 6 will be reduced by all monies that Company is obligated to pay those producers in connection with that project.

(c) At your written request pursuant to Company's standard letter of direction executed by you and actually received by Company within sixty (60) days after the initial United States release of the Album concerned, Company will pay a royalty to any mutually approved independent third party producer engaged by you (or by Company, on your behalf) and to whom you are obligated to pay a royalty (the "Producing Royalty") in respect of Net Sales of Records (and other exploitations of the applicable Masters) released hereunder. The Producing Royalty will be computed, adjusted and paid in the same manner, at the same time and subject to the same conditions as the royalty payable to you, but at a basic rate of no more than three percent (3%), with escalations in such basic rate at certain sales levels, without your prior written consent, with proportionate reductions on all sales (and other exploitations of the applicable Masters) for which reduced royalties are payable under this agreement. The Producing Royalty will not be payable to the producer concerned until Company has recouped (pursuant to the terms hereof) all Recording Costs attributable to the Recordings concerned. Such recoupment will be computed at your net royalty rate (as reduced to reflect the deduction of the Producing Royalty and royalties payable to all other third party royalty participants). After such recoupment, the Producing Royalty will be computed retroactively and paid (as provided above) on the Net Sales of the Record concerned from the first such Record sold (after recoupment of the applicable producer advance). The amount of the Producing Royalty will be recoupable by Company from all monies (excluding Mechanical Royalties) payable or becoming payable to you hereunder. Company's compliance with your request to pay any such Producing Royalty will not constitute the producer, or any payee on behalf of the producer, a beneficiary of or a party to this agreement. All Producing Royalty payments hereunder will constitute payments to you, and Company will have no liability by reason of any such payment, or failure to make same. You hereby indemnify and hold Company harmless (pursuant to the terms of paragraph 13.05 below) against any claims asserted against Company in connection with any such Producing Royalty.

3.05. In connection with each recording session conducted hereunder, you will comply with the following procedures required by United States immigration law:

(a) Before any individual renders services in connection with the recording of any Master hereunder (including, without limitation, each background instrumentalist, background vocalist, producer and engineer):

   (i) You will require each such individual to complete and sign the EMPLOYEE INFORMATION AND VERIFICATION ("employee section") of a U.S. Immigration and Naturalization Service ("INS") Employment Eligibility Certificate ("Form I-9"), unless you have already obtained (and retained) such certificate from that individual within the past three (3) years;

   (ii) You will complete and sign the EMPLOYER REVIEW AND VERIFICATION ("employer section") of each such certificate; and

   (iii) You will attach copies of the documents establishing identity and employment eligibility that you examine in accordance with the instructions in the employer section.

(b) You will not permit any such Person who fails to complete the employee section (or to furnish you with the required documentation) to render any services in connection with Recordings made under this agreement.

(c) You will deliver the employee and employer certificates (with copies of the necessary documents attached) to Company within seventy-two (72) hours after the conclusion of the session concerned.

(d) You will comply with any revised or additional verification and documentation procedures required by the INS in the future.

4. *Recoupable Costs*

   4.01. Company may pay from the applicable Authorized Budget all union scale payments required to be paid to Artist in connection with Masters made hereunder, all costs of instrumental, vocal and other personnel specifically approved by Company for the recording of such Masters, and all other amounts required to be paid by Company pursuant to any applicable law or any collective bargaining agreement between Company and any union representing persons who render services in connection with such Masters. Notwithstanding the foregoing, you agree that the Advances hereunder include the prepayment of session union scale as provided in the applicable union codes, and you agree to complete any documentation required by the applicable union to implement this sentence. At Company's request, union contracts will be filed by you and supplied to Company and pension benefits will be paid on your behalf from the Authorized Budget, which payments will be an Advance.

   4.02. All Recording Costs hereunder and all costs associated with the creation of Artwork (including, without limitation, art, photos, graphic design, etc.) constitute Advances. In addition, all packaging costs in excess of Company's then standard design, engraving or manufacturing costs with respect to a standard Record package are your responsibility and may, at Company's election, be recovered as provided in paragraph 5.03 below. Company will notify you if Company anticipates incurring any such excess design, engraving or manufacturing costs in connection with any Album package hereunder. Provided that changing the Album package would not unreasonably delay the scheduled release of the Record concerned, if you object to such excess design, engraving or manufacturing costs in connection with an Album package not prepared by you or not prepared at your request, and such written objection is received by Company within five (5) days after Company's notice, and Company nevertheless uses such package, Company will not charge you for the excess. No failure to so notify you will be deemed a breach hereof; provided that, if Company fails to so notify you, Company will not charge you for any such excess. One half (1/2) of all Video Costs will be recoupable from audio Record royalties. To the extent that any Video Costs are not recoupable or recouped from audio Record royalties, such costs will be recoupable from monies (other than Mechanical Royalties) otherwise payable to you from the exploitation of Videos hereunder. Fifty percent (50%) of all costs paid by Company in connection with any television campaign in the United States in conjunction with Records featuring your performance will constitute Advances. All costs or expenses paid by Company to third parties in connection with independent publicity will constitute Advances. Fifty percent (50%) of all costs incurred by Company in connection with independent marketing and/or independent promotion of Records featuring your performance will constitute Advances. All costs paid by Company in connection with "live" public performances by you and/or to purchase articles of dress and styling such as clothing, shoes, jewelry, makeup, hair and body styling will constitute Advances. All costs specifically with respect to you incurred in connection with creating the so-called "enhanced" or multimedia portion of an enhanced CD, CD Plus, CD ROM, DVD, or any other similar configuration (whether now known or hereafter created) embodying Masters hereunder (the "Enhanced Costs") including, without limitation, ECD Material, will constitute Advances. All costs paid or incurred by Company in connection with the creation, development and maintenance of any website featuring the promotion of you and Records embodying Masters will constitute Advances.

4.03. The portion of the Recording Costs incurred in the making of a Joint Recording to be charged as an Advance will be computed by multiplying the aggregate amount of total Recording Costs incurred in making that Joint Recording by the same fraction used in determining the royalties payable to you in respect of that Joint Recording.

5. *Additional Advances*

    5.01. All monies paid to you during the term of this agreement, as well as all monies paid on your behalf at your request, or as otherwise contemplated herein, other than royalties paid pursuant to this agreement, constitute Advances unless otherwise expressly agreed to in writing by Company, or except as otherwise set forth in this agreement.

    5.02. (a) In connection with your Delivery to Company of the Masters constituting each Album Delivered in fulfillment of your Recording Commitment, Company will pay you an Advance in the amounts indicated below:

        (i) With respect to the First Album, $ _____ .

        (ii) With respect to each subsequent Album, recorded and delivered hereunder in fulfillment of your Recording Commitment, the Formula Amount, but not less than the minimum nor more than the maximum amount set forth below:

Album	Minimum/Maximum
Second	$ _____ /$ _____
Third	$ _____ /$ _____
Fourth	$ _____ /$ _____
Fifth	$ _____ /$ _____
Sixth	$ _____ /$ _____
Seventh	$ _____ /$ _____

        (iii) The "Formula Amount" for a particular Album recorded and delivered hereunder in fulfillment of your Recording Commitment will mean an amount equal to twenty percent (20%) of whichever of the following amounts is less: (A) the amount of the royalties, after the retention of reserves (which, for purposes of this calculation, will be fifteen percent (15%)), earned by you hereunder from Net Sales (defined below) of the immediately preceding Album delivered hereunder in fulfillment of your Recording Commitment; or (B) the average of the amounts of such royalties so earned by you hereunder on the two (2) immediately preceding Albums delivered hereunder in fulfillment of your Recording Commitment. In either case, the amount of royalties with respect to any preceding Album will be computed as of the end of the month in which occurs the date which is twelve (12) months following the initial commercial release in the United States of the preceding Album concerned.

(b) The Advances referred to in this paragraph 5.02 will be paid one-half (1/2) upon commencement of recording of the Album concerned (or, in the case of the First Album, upon execution of this agreement) and the balance upon Delivery of such Album.

5.03. If the Recording Costs and other Advances paid or reimbursed by Company for any Recording in fulfillment of your Recording Commitment exceed the Authorized Budget therefor, you will be solely responsible for such excess, it being agreed that if Company elects to pay such excess, such payment will be a direct debt from you to Company which, in addition to any other available remedies, Company may recover from any sums payable to you or your designees. Notwithstanding the foregoing, if the excess was caused solely by Company, Company shall be responsible for such excess, however, Company may recover the amount of such excess from royalties (other than Mechanical Royalties) payable to you or your designees.

6. *Royalties*

   6.01. In consideration of the copyright ownership provided below, Company's rights to use your name and likeness as provided herein, and the other agreements, representations and warranties contained herein, Company agrees to pay you in connection with the Net Sale of Records consisting entirely of Masters hereunder and sold by Company or its licensees and in connection with other commercial exploitations of the Masters by Company or its licensees, a royalty computed at the applicable percentage indicated below, of the applicable Royalty Base Price with respect to the Record concerned, it being agreed that such royalties will be computed and paid in accordance with Article 7 below and the other provisions set forth herein.

   (a) Subject to the other provisions of this Article 6, the basic rate on USNRC Net Sales of Albums ("Basic Album Rate"): ____%

   (b) Subject to the other provisions of this Article 6, the basic rate on USNRC Net Sales of Singles (the "Basic Singles Rate"): ____%

   (c) The Basic Album Rate and the Basic Single Rate, as applicable, is sometimes referred to below as the "Basic Rate."

   (d) On Records sold for distribution Through Normal Retail Channels outside the United States, the Basic Rate for the country concerned shall be one-half (1/2) of the percentage of the United States Basic Rate; provided, however, in no instance shall you receive more than fifty percent (50%) of Company's Net Receipts therefrom.

   6.02 With respect to Midline Records, EPs, Records sold through a Developing Artist Series at a price equivalent to a Midline Record, and Records sold to the armed forces post exchanges, the royalty rate will be two thirds (2/3) of the Basic Rate in the country concerned for the configuration concerned. With respect to Budget Records, so-called "premium" Records, Records sold through a Developing Artist Series at a price equivalent to a Budget Record, and Records sold to the United States or a state or local government or sold by Company itself in the United States in connection with a direct response television campaign, direct mail or mail order, the royalty rate will be one-half (1/2) of the Basic Rate in the country concerned for the configuration concerned. With respect to any Record sold outside the United States by Company's licensees in conjunction with a television advertising campaign, in the event Company's royalties are reduced in connection therewith, your royalties shall be likewise proportionately reduced. With respect to any Multiple Record Album, the royalty rate will be the Basic Rate in the country concerned for the configuration concerned if, at the beginning of the royalty accounting period concerned, the Suggested Retail List Price of such Album is at least the number of cassettes, compact discs or other configuration packaged together times the Suggested Retail List Price for "top-line" Albums marketed by Company or its principal licensee in the country where the Multiple Record Album is sold (the "top-line" price). If the Suggested Retail List Price applicable to such Multiple Record Album is less than the number of cassettes, compact discs or other configuration packaged together times the "top-line" price, then the applicable royalty rate for such Multiple Record Album will be equal to the otherwise applicable royalty rate multiplied by a fraction, the numerator of which is the Suggested Retail List Price of such Multiple Record Album, and the denominator of which is the number of cassettes, compact discs or other configuration packaged together times the "top-line" price (but not less than one half (1/2) of the applicable royalty rate prescribed in paragraph 6.01 for such Album). With respect to Records in the compact disc configuration, the royalty rate (which will be deemed to be the Basic Album Rate or Basic Singles Rate, as applicable, with respect to such configuration), will be eighty-five percent (85%) of the otherwise applicable royalty rate in the applicable country for the configuration and price category concerned. With respect to Electronic Transmissions, the royalty rate (which will be deemed to be the Basic Album Rate or Basic Singles Rate, as applicable, with respect to such configuration) will be one hundred percent (100%) of the otherwise applicable royalty rate in the applicable country for the configuration and price category concerned. With respect to Records sold in the form of new configurations (including, but not limited to, Digital Compact Cassette, Mini Disc, DVD Audio, and audiophile Records), the royalty rate (which shall be deemed to be the Basic Album Rate or Basic Singles Rate, as applicable, with respect to such configurations) will be seventy-five percent (75%) of the otherwise applicable royalty rate in the applicable country for the configuration and price category concerned. Notwithstanding the foregoing, with respect to each particular new configuration sold three (3) years after its initial commercial release by Company hereunder, upon your request Company shall negotiate with you in good faith in accordance with then current industry practices and the rates then being paid in the industry for artists of your stature for the purpose and intent of attempting to agree upon the percentage royalty rate payable to you prospectively with respect to such particular new configuration; however, in no event shall such percentage royalty rate exceed one hundred percent (100%) of the otherwise applicable royalty rate in the applicable country for the configuration and price category concerned; provided, further, Company will have the right to continue to exploit Records in such new configuration pending such negotiations.

6.03. (a) Your royalty will be the sum equal to fifty percent (50%) of Company's Net Receipts with respect to the following Records and/or exploitation of Masters recorded hereunder: (i) Records sold through record clubs or similar sales plans; (ii) licenses for methods of distribution such as "key outlet marketing" (distribution through retail fulfillment centers in conjunction with special advertisements on radio or television), direct mail, mail order, or by any combination of the methods set forth above or other methods; (iii) licenses for distribution other than Through Normal Retail Channels or other than by the primary Distributor(s) of Company's Records in the territory concerned for the configuration concerned; and (iv) licenses for use of the Masters produced hereunder for which a royalty is not specifically set forth herein.

(b) In respect of any Ancillary Exploitation for which Company receives a royalty or other payment which is directly and solely attributable to such Ancillary Exploitation, your royalty will be an amount equal to one-half (1/2) of Company's Net Receipts solely derived from such Ancillary Exploitation.

6.04. (a) If Company licenses Videos produced hereunder, your royalty will be one half (1/2) of Company's Net Receipts derived therefrom after deducting from gross receipts a fee, in lieu of any overhead or distribution fee, of twenty percent (20%) of the gross receipts in connection therewith. It is specifically agreed that Company has and will have the right to license Videos to third parties (e.g., club services) for no payment, in which case no payment will be made to you in connection therewith.

(b) With respect to home video devices embodying Videos produced hereunder manufactured and distributed by Company or its exclusive licensee in the country concerned, you will be entitled to a royalty computed as provided in this Article, but the following rates will apply instead of the rates specified in paragraph 6.01 above: (i) On units sold for distribution in the United States: 15% of the applicable Royalty Base Price for home video devices provided

Company's wholesale price is Ten Dollars ($10.00) or less; seventeen and one half percent (17 1/2%) of the applicable Royalty Base Price for home video devices provided Company's wholesale price is greater than Ten Dollars ($10.00) but less than or equal to Fourteen Dollars ($14.00); and twenty percent (20%) of the applicable Royalty Base Price for home video devices provided Company's wholesale price is in excess of Fourteen Dollars ($14.00); and (ii) On units sold for distribution outside the United States: 10% of the applicable Royalty Base Price. Said royalties are inclusive of any third party payments required in connection with the sale of such devices including, without limitation, artist and producer royalties and copyright payments.

6.05. As to a Record not consisting entirely of Masters recorded hereunder or Videos produced hereunder, the otherwise applicable royalty rate will be prorated on the basis of the number of Masters recorded hereunder or Videos produced hereunder embodied on such Record compared to the total number of Masters or Videos (including the Masters recorded hereunder and Videos produced hereunder) contained on such Record. As to a Record consisting of Masters recorded hereunder where the various Masters bear different royalty rates, the otherwise applicable royalty rate for each Master will be prorated on the basis of the total number of Masters embodied on such Record. As to Joint Recordings, the royalty rate will be the royalty rate provided for herein divided by the number of Persons with respect to whom Company is obligated to pay a royalty (including you). For purposes of this paragraph 6.05 a group that typically records as a group will be deemed to be one (1) Person, provided that such group is paid the same royalty as would otherwise be payable if one (1) individual had rendered the same services.

6.06. No royalties will be due or payable in respect of (a) Records furnished on a no-charge basis or sold to disc jockeys, publishers, employees of Company or its licensees, motion picture companies, radio and television stations and other customary recipients of free Records, or discounted or promotional Records sold for less than or equal to fifty percent (50%) of the Record's highest posted wholesale list price; (b) Records sold at close-out or "cut-out" prices, as surplus, for scrap, at less than inventory cost, or at a discount of seventy-five percent (75%) or more off their SRLP (whether or not such Records would otherwise constitute Budget Records and/or are intended for resale to third parties); (c) Records (or fractions thereof) given away or shipped on a so-called "no charge", "freebie" or "bonus" basis (whether or not intended for resale to third parties); (d) Records sold, distributed or furnished on a no-charge basis to members, applicants or other participants in any "record club"; (e) Records (other than Singles) sold by Company or Company's Distributor (or their respective licensees) as "samplers" and/or directly to consumers for a price of $3.00 per unit or less, as well as Singles sold in the form of "picture discs" not in excess of 25,000 Net Sales per such Single; and (f) Records sold at a discount from the Record's posted wholesale list price (but for more than fifty percent (50%) of such price). In determining the number of Records as to which no royalties are payable pursuant to the preceding clause (f), Company will multiply the percentage amount of such discount by the number of Records sold at such discount. So-called standard "free goods" on Albums distributed in the United States, if any, will not exceed twenty percent (20%) of all Albums distributed unless Company pays you on any such excess.

6.07. Company may at some time change the method by which it computes royalties in the United States from a retail basis to some other basis (the "New Basis"), such as, without limitation, a wholesale basis. The New Basis will replace the then-current Royalty Base Price and the royalty rates will be adjusted to the appropriate royalty which, when applied to the New Basis, will yield the same dollars-and-cents royalty amounts payable with respect to the Record concerned as was payable immediately prior to the change to the New Basis. If a Record was not theretofore sold in a particular configuration or at a particular price (e.g., a Budget Record), the adjusted royalty rate for any such configuration will be the adjusted royalty rate on top-line Albums multiplied by a fraction, the numerator of which is the royalty rate for sales in the configuration concerned prior to the New Basis and the denominator of which is the royalty rate for sales of top-line Albums prior to the New Basis. If there are other adjustments made by Company that would otherwise make the New Basis more favorable (a particular example of which might be the distribution of smaller quantities of free goods than theretofore distributed), then the benefits of such other adjustments will be taken into consideration in adjusting the royalty rate.

6.08. The royalty payable to you hereunder includes all royalties due you, the individual producers and all other Persons in connection with the sale of Records or other exploitation of Masters made hereunder, excluding Mechanical Royalties and union "per record" royalties.

7. *Royalty Accountings*

7.01. Company will compute your royalties as of each June 30 and December 31 for the prior six (6) months, in respect of each such six (6) month period in which there are sales or returns of Records or other exploitations of Masters on which royalties are payable to you. On or before the next September 30 with respect to the period ending June 30, and on or before March 31 with respect to the period ending December 31, Company will send you a statement covering those royalties and will remit to you the net amount of such royalties, if any, after deducting any and all unrecouped Advances and chargeable costs under this agreement and such amount, if any, that Company may be required to withhold pursuant to the applicable state tax laws, the U.S. Tax Regulations, or any other applicable statute, regulation, treaty, or law. After the term hereof, no royalty statements will be required for periods during which no additional royalties accrue. In computing the number of Records sold, only Records for which Company has been paid or has received final credit against a prior advance received by Company in the United States will be deemed sold, and Company will have the right to deduct returns and credits of any nature and to withhold reasonable reserves therefor from payments otherwise due you, not to exceed thirty-five percent (35%). Company will liquidate any such reserves within four (4) full accounting periods after the period in which such reserves were initially established. If Company makes any overpayment to you (e.g., by reason of an accounting error or by paying royalties on Records returned later), you will reimburse Company to the extent Company does not deduct such sums from monies due you hereunder.

7.02. Royalties for Records sold for distribution outside the United States ("foreign sales") will be computed in the same national currency and at the same rate of exchange as Company is accounted to by its licensees with respect to the sale concerned and will be subject to costs of conversion and any taxes applicable to royalties remitted by or received from foreign sources. Royalties on Records sold outside the United States are not due and payable by Company until payment or credit therefor has been received by Company in the United States in United States dollars. For purposes of accounting to you, Company will treat any foreign sale as a sale made during the same six (6) month period in which Company receives its licensee's accounting and payment for that sale. If Company does not receive payment in the United States in United States Dollars and is required to accept payment in foreign currency or in a foreign country, Company will notify you thereof and deposit to your credit (at your request and expense) in such currency in a depository selected by you in the country in which Company accepts payment your share of royalties due and payable with respect to such sales. Such deposit will fulfill Company's obligations in connection therewith. If any law, government ruling or other restriction affects the amount that an Company licensee can remit to Company, Company may deduct from your royalties an amount proportionate to the reduction in such licensee's remittances.

7.03. All royalty statements rendered by Company will be conclusively binding upon you and not subject to any objection by you for any reason unless specific objection in writing, stating the basis thereof, is given to Company within two (2) years from the date such statement is rendered and an audit pursuant to paragraph 7.04 for that statement is completed within three (3) months after such objection notice is given, unless the audit could not be completed within such three (3) month period solely as a result of the acts or omissions of Company. Failure to make such written objection or conduct the audit within said time periods will be deemed to be your approval of such statement, your waiver of such audit rights, and your waiver of the right to sue Company for additional royalties in connection with the applicable accounting period. Each statement will be deemed rendered when due unless you notify Company that the applicable statement was not received by you and such notice is given within sixty (60) days after the applicable due date specified in paragraph 7.01 above, in which event the statement will be deemed rendered on the date actually sent by Company. You will not have the right to sue Company in connection with any royalty accounting, or to sue Company for monies due on account of the exploitation of Masters hereunder during the period a royalty accounting covers, whether from the sale of Records or otherwise, unless you commence the suit within the earlier of twelve (12) months after commencement of your audit for the applicable period or three (3) years from the date such statement is rendered for the applicable period.

7.04. You may, at your own expense, audit Company's books and records directly relating to this agreement that report the sales or other exploitation of Records or Ancillary Exploitations for which royalties or other monies are payable hereunder. You may make such audit only for the purpose of verifying the accuracy of statements sent to you hereunder and only as provided herein. You may initiate such audit only by giving notice to Company at least thirty (30) days prior to the date you intend to commence your audit. Your audit will be conducted by a reputable independent certified public accountant experienced in recording industry audits in such a manner so as not to disrupt Company's other functions and will be completed promptly. You may audit a particular statement only once and only within two (2) years after the date such statement is rendered as provided in paragraph 7.03 above, unless the audit was delayed by Company in which event such two (2) year period shall be extended by the number of days of the delay caused by Company. Your audit may be conducted only during Company's usual business hours and at the place where it keeps the books and records to be examined. You will not be entitled to examine any manufacturing records or any other records that do not specifically report sales of Records or free distribution of Records on which royalties are payable hereunder. Your auditor will review his tentative written findings with a member of Company's finance staff designated by Company before rendering a report to you so as to remedy any factual errors and clarify any issues that may have resulted from misunderstanding.

8. *Company's Additional Rights*

8.01. You warrant, represent and agree that throughout the Territory Company is the sole, exclusive and perpetual owner of all Masters Delivered hereunder or otherwise recorded by you during the term of this agreement, all so-called "demonstration" recordings recorded by you prior to or during the Term (including, without limitation, any and all Master Recordings embodying your performances that were previously submitted to Company for Company's review), all Videos embodying those Masters or otherwise produced hereunder, and all artwork created for use in connection with the Masters, Videos, and/or ECD Material (individually and collectively referred to herein as "Artwork"), which ownership entitles Company, among other things, to all right, title and interest in the copyright in and to the Masters, Videos (but excluding the copyrights in the Compositions contained in the Masters and Videos) and Artwork. Each Master, Video and Artwork made under this agreement or during its term, from inception, will be considered a "work made for hire" for Company; if any such Master, Video or Artwork is determined not to be such a "work," it will be deemed transferred to Company by this agreement, together with all rights and title in and to it. You warrant, represent and agree that all Masters and Videos (but excluding the underlying Compositions embodied in the Masters and Videos) made under this agreement or during its term (including duplicates, work tapes, etc.), the performances contained thereon and the Recordings and Records derived therefrom and the related Artwork, from the inception of their creation, are the sole property of Company, in perpetuity, free from any claims by you or any other Person, and Company has the right to use and control same subject to the terms herein. Company (or Company's designees) has the exclusive right to copyright all such Masters, Videos and Artwork in its name as the author and owner of them and to secure any and all renewals and extensions of such copyright throughout the Territory. You will execute and deliver to Company such instruments of transfer and other documents submitted to you by Company regarding the rights of Company or its designees in the Masters, Videos and Artwork subject to this agreement as Company may reasonably request to carry out the purposes of this agreement, and Company may sign such documents in your name (and you hereby appoint Company your agent and attorney-in-fact for such purposes) and make appropriate disposition of them consistent with this agreement. Company shall give you five (5) business days notice before signing in your name, any document submitted to you by Company, provided Company may dispense with that waiting period when necessary, in Company's good faith judgment, to protect or enforce Company's rights. As a non-material obligation, Company shall provide you with copies of documents signed by Company in your name. Company will have the right to include on any Artwork hereunder all such, trademarks, trade names, information, logos and other items, as Company customarily includes on such Artwork, as applicable, including, without limitation, Internet Addresses, so-called "watermarks," "meta-data," and "hyperlinks" to Internet Addresses.

8.02. Without limiting the generality of the foregoing, and except as otherwise specifically set forth in this agreement, Company and any Person authorized by Company has the unlimited and exclusive rights to manufacture and/or distribute Records by any and all methods now or hereafter known embodying any portion or all of the performances embodied on Masters hereunder; to publicly perform such Records and to permit the public performance thereof in any medium now or hereafter known; to import, export, sell, transfer, transmit, lease, rent, deal in or otherwise dispose of such Masters (including without limitation, by way of Electronic Transmission) and Records derived therefrom throughout the Territory under any trademarks, trade names or labels designated by Company; to remix, edit or adapt the Masters to conform to technological or commercial requirements in various formats now or hereafter known or developed, or to eliminate material which might subject Company to and legal action; to use and authorize the use of the Masters for background music, synchronization in motion pictures and television soundtracks and other similar purposes, including, without limitation, use on transportation and in commercials for any product in any and all media, without any payment other than as provided herein; or Company and its subsidiaries, affiliates and licensees may, at their election, delay or refrain from doing any one or more of the foregoing.

8.03. Company and any licensee of Company each has the perpetual right, without liability to any Person, and may grant to others the right, to reproduce, print, publish or disseminate in any medium your name, portrait, picture and likeness and the name, portrait, picture and likeness of any individual producer and all other Persons performing services in connection with Masters made under this agreement (including, without limitation, all professional, group and other assumed or fictitious names used by you and them), and biographical material concerning you and them solely for purposes of advertising, promotion and trade in connection with you, the making and exploitation of Records hereunder on Websites, in Website Material and ECD Material, and for advertising and for purposes of trade in connection therewith, and general goodwill advertising of Company and its Record business. Notwithstanding the foregoing, in the event there are contractual restrictions, which are customary in the Record business, in connection with the use of any producer's or other Person's portraits, pictures, likenesses or biographical material, Company shall abide by such restrictions; provided that, the relevant portions of such contracts containing such restrictions are provided to Company not later than the Delivery of the applicable Masters. The uses authorized by the preceding sentence include, without limitation, the use of those names, portraits, pictures, and likenesses in the marketing of Records. During the term hereof, Company and its licensees may, in the Territory, bill, advertise and describe you as an exclusive Company artist or by a similar designation. Subject to your prior professional commitments, you will from time to time appear for photography, poster and cover art and the like, under the reasonable direction of Company or its nominees, appear for on-line "chats" hosted on Websites, including, without limitation, the Company Artist Website, and for interviews with representatives of the media and Company's publicity personnel, and appear and perform at promotional events such as so-called "in-store" performances. You will not be entitled to any compensation for such services, except as may be required by applicable union agreements; provided, however, that if you are required to travel outside of a fifty (50) mile radius of your then place of residence, Company will reimburse you on a non-recoupable basis for the reasonable travel and living expenses incurred by you in connection with the rendition of services at Company's direction pursuant to a budget approved in advance by Company in writing.

8.04. (a) Without limiting the generality of the rights granted to Company pursuant to this agreement, upon at least thirty (30) days prior written notice to you, Company and any licensee of Company each has the perpetual and exclusive right and may grant to others the right, without liability to any Person, to: (1) create, maintain and host one (1) Company Artist Website; and/or (2) register and use the Artist Domain Name as the Internet Address in connection with such Company Artist Website. The Artist Domain Name, and all rights thereto or derived therefrom shall be Company's property throughout the Territory and in perpetuity subject to the terms and conditions contained in this agreement. During the term hereof, you will upon Company's request link your "official" Website to the Company Artist Website and Company's "official" Website, and Company will upon your request link the Company Artist Website to your "official" Website; provided, however, Company shall have the right to discontinue such link if in its reasonable judgment, your "official" site contains material which may constitute defamation, libel, violate or infringe upon any right, including, without limitation, the right to privacy, of any Person, or might subject Company to any civil or criminal action, or may adversely impact Company.

(b) The Company Artist Website, Company Website Material and ECD Material will be deemed a Material as provided herein. Company will have the rights in and to the Company Artist Website, Company Website Material and ECD Material as are otherwise applicable hereto with respect to Masters made hereunder. Without limiting the generality of the foregoing, and except with respect to Material supplied by you but owned by an unrelated third party, Company is and will be the sole owner of all worldwide rights in and to the Company Artist Website and all ECD Material and Company Website Material created hereunder, all individual elements thereof, and the selection and arrangement of such elements, including the worldwide copyrights therein and thereto, throughout the Territory and in perpetuity. You will notify Company of any Material owned by a third party contemporaneously with you supplying such Material to Company.

8.05. If, at any time during the term hereof, you shall commit any act or become involved in a situation or occurrence which brings you and/ or Company in public disrepute, contempt, scandal or ridicule or which tends to shock, insult or offend the community at large or any substantial group or class thereof, or which reflects unfavorably upon Company's reputation, then Company shall have the right to terminate the term of this agreement without any further obligation to you, except to pay such sums as may have become due under the terms of this agreement prior to such commission or involvement.

9. *Videos: Marketing and Miscellaneous Restrictions*

9.01. (a) With respect to Videos made hereunder during the term hereof, the selection(s) to be embodied in each Video will be mutually designated by you and Company, provided, however (i) in the event of a disagreement, Company shall have the final say, and (ii) you will be deemed to have approved any selection that has been or will be embodied on a Single.

(b) Each Video will be shot on a date or dates and at a location or locations to be designated by Company, subject to your prior professional commitments.

(c) The producer, director, and concept or script for each Video will be approved by both you and Company. Company will engage the producer, director and other production personnel for each Video and will pay the production costs of each Video in an amount not in excess of a budget to be established in advance by Company (the "Production Budget"). You will pay any and all production costs for each Video in excess of the Production Budget where such excess is caused by your acts or omissions, provided that if such excess is not caused by your acts or omissions, such excess will be an expense in connection with such Video, recoupable in the same manner as costs incurred in connection with the Production Budget. In the event that Company pays any production costs that are your responsibility pursuant to the foregoing (which Company is in no way obligated to do), you will promptly reimburse Company for such excess upon demand and, without limiting Company's other rights and remedies, Company may deduct an amount equal to such excess from any monies (other than Mechanical Royalties) otherwise payable to you hereunder. Your compensation for performing in each Video (as opposed to your compensation with respect to the exploitation of such Videos, which is provided elsewhere herein) will be limited to any minimum amounts required to be paid for such performances pursuant to any collective bargaining agreements pertaining thereto, provided, however, that you hereby waive any right to receive such compensation to the extent such right may be waived.

(d) Company is and will be the sole owner of all worldwide rights in and to each Video (including the worldwide copyrights therein and thereto, but not the underlying musical composition embodied therein).

(e) You will issue (or cause the music publishing companies having the right to do so to issue) (1) worldwide, perpetual synchronization licenses, and (2) perpetual licenses for public performance in the United States (to the extent that ASCAP, BMI or SESAC are unable to issue same), to Company at no cost for the use of all Controlled Compositions in any Video effective as of the commencement of production of the applicable Video (and your execution of this agreement constitutes the issuance of such licenses by any music publishing company that is owned or controlled by you or any Person owned or controlled by you). In the event that you fail to cause any such music publishing company to issue any such license to Company, or if Company is required to pay any fee to such music publishing company in order to obtain any such license, Company will have the right to deduct the amount of such license fee from any and all sums otherwise payable to you hereunder. Notwithstanding the foregoing, although the synchronization license is perpetual and remains in effect, if the cost incurred with any such Video is entirely recouped, then after such recoupment, and only with respect to prospective commercial uses of such Video, Company and you will negotiate in good faith with respect to compensation consistent with the then-current music industry standards, to be paid by Company for such a synchronization license for the Controlled Compositions used in such Video.

(f) Company will have the right to use and allow others to use each Video for Advertising and Promotional Purposes and for Commercial Purposes.

(g) Each Video will be deemed a Material as provided herein. Company will have the rights in and to each Video as are otherwise applicable hereto with respect to Masters made hereunder, including, without limitation, the right to use and publish, and to permit others to use and publish, your and Artist's name and likeness in each Video and for advertising and purposes of trade in connection therewith.

(h) Company is under no obligation whatsoever to produce Videos hereunder.

9.02. (a) Provided you have fulfilled all your material obligations under this agreement, Company will commercially release in the United States each Album recorded in fulfillment of your Recording Commitment within six (6) months after Delivery of the Album concerned. If Company fails to do so you may notify within sixty (60) days after the end of the six (6) month period concerned, that you intend to terminate the term of this agreement unless Company commercially releases the Album within two (2) months after Company's receipt of your notice (the "cure period"). If Company fails to commercially release the Album before the end of the cure period, you may terminate the term of this agreement by giving Company notice (the "Termination Notice") within thirty (30) days after the end of the cure period. On receipt by Company of your Termination Notice, the term of this agreement will end and all parties will be deemed to have fulfilled their obligations hereunder except those obligations that survive the end of the term (e.g., warranties, re-recording restrictions and obligations to pay royalties). Notwithstanding paragraph 1.01, in the event you fail to give Company the Termination Notice within said thirty (30) day period with respect to the last Album to be Delivered in fulfillment of your Recording Commitment in any Contract Period, you may at any time thereafter notify Company that the applicable Contract Period will end on the date six (6) months from the date of the notice (but in no event earlier than one (1) year from the commencement of the Period), and Company will have through the last day of the Contract Period to exercise its Option (if any) for the next Option Period. Your only remedy for failure by Company to release an Album will be as described in this paragraph. If you fail to give Company the Termination Notice within the period specified, your right to terminate as to that Album will lapse.

(b) The running of each of the six (6) month and two (2) month periods referred to in subparagraph 9.02(a) will be suspended (and the expiration date of each of those periods will be postponed) for the period of any suspension of the running of the term of this agreement. If any such six (6) month or two (2) month period would otherwise expire on a date between November 15 and the next January 16, its running will be suspended for the duration of the period between November 15 and January 16 and its expiration date will be postponed by the same amount of time [i.e., sixty-two (62) days].

9.03. Within a reasonable time following the execution hereof, you may supply Company with six (6) approved pictures and approved biographical material to be used by Company pursuant to paragraph 8.03 above. In the event that Company reasonably disapproves of the pictures or biographical material supplied by you, Company will make available to you for your approval pictures and biographical material concerning you to be so used by Company. Your approval will not be unreasonably withheld and will be deemed given unless your notice of disapproval (including the reason) has been received by Company within ten (10) business days after the material has been made available to you. In the event that you timely disapprove of any pictures or biographical material, you will, within seven (7) days of the date of your disapproval notice, supply to Company approved pictures or biographical materials for use by Company hereunder. In the event that the pictures and/or biographical materials supplied by you, pursuant to the preceding sentence, are not satisfactory to Company in its good-faith opinion, or in the event that you do not supply the pictures or biographical materials to Company pursuant to this paragraph, Company will have the right to select and use such pictures or biographical materials as its determines in its sole discretion, and you will have no approval rights in respect thereof. In connection with each new Album hereunder, you may supply Company with more recently approved pictures and a more recent biography for use by Company hereunder. If you do so, the procedures set forth in this paragraph will apply. No inadvertent failure by Company to comply with this paragraph will constitute a breach of this agreement. Your sole remedy for any failure by Company to comply with this paragraph will be prospective cure with respect to materials prepared after the notice period specified in paragraph 16.06 below.

9.04. (a) With respect to the Company Artist Website, Company Website Material and ECD Material, all artwork and other creative elements produced in connection therewith, including, without limitation, production personnel, shall be mutually approved by you and Company, provided, however, (i) in the event of a disagreement, Company shall have the final say; (ii) any Artwork or other materials furnished or approved by you for any other purpose hereunder shall be deemed approved by you for use in connection with the Company Artist Website, Company Website Material and ECD Material, and provided further that Company will have the right to include on the Company Artist Website and within ECD Material so-called "hyperlinks" to Artist Domain Names and other URL's in Company's sole discretion.

(b) You will issue (or cause the music publishing companies having the right to do so to issue) (1) worldwide, perpetual synchronization licenses, and (2) worldwide, perpetual licenses for public performance to Company at no cost for the promotional use of all Controlled Compositions in Website Material and ECD Material effective as of the commencement of production of the applicable Website Material or ECD Material (and your execution of this agreement constitutes the issuance of such licenses by you and any music publishing company that is owned or controlled by you or any Person owned or controlled by you). In the event that you fail to cause any such music publishing company to issue any such license to Company, or if Company is required to pay any fee to such music publishing company in order to obtain any such license, Company will have the right to deduct the amount of such license fee from any and all sums otherwise payable to you hereunder.

(c) Company will have the right to use and allow others to use Website Material and ECD Material for Advertising and Promotional Purposes and for Commercial Purposes.

(d) You will exert reasonable efforts to supply Company, at Company's request, with Website Material for possible inclusion on the Company Artist Website, including, without limitation transcripts of all of your published interviews, transcripts of all articles relating to you, photographs, and other similar materials.

9.05. You may use Record packaging artwork for non-Record merchandising purposes, upon payment to Company of fifty percent (50%) of the out-of-pocket costs paid or incurred by Company in the creation thereof, and agreement to any reasonable conditions imposed by Company (e.g., copyright notices). Company will provide said artwork to you without warranty or representation, express or implied, you shall be responsible for obtaining any third party clearances (e.g., photographers) and you will indemnify Company, in the manner provided in paragraph 13.05 hereof, against any and all claims, damages, liabilities, costs and expenses, including reasonable counsel fees, arising out of any use of said artwork or exercise such rights by you or any Person deriving rights from you.

10. *Licenses for Musical Compositions*

10.01.(a) (i) You hereby grant to Company and its designees an irrevocable license under copyright to reproduce each Controlled Composition on Records and distribute such Records in the United States and Canada.

(ii) For that license, Company and its designees will pay you or your designee Mechanical Royalties, on the basis of Net Sales, at the following rate (the "Controlled Rate"):

(A) *On audio Phonograph Records distributed in the United States*: If the copyright law of the United States provides for a minimum compulsory rate: The rate equal to one hundred percent (100%) of "United States Statutory Rate." The term "United States Statutory Rate" shall mean the minimum compulsory license rate applicable to the use of musical compositions on audio Phonograph Records under the United States copyright law at the time of Delivery of the Master concerned but in no event later than the last date for timely Delivery of such Master pursuant to Article 2. (The U.S. minimum compulsory rate is $.085 per Composition as of January 1, 2004). If no such mechanical royalty rate exists, however, the "United States Statutory Rate" shall mean the minimum license rate agreed to by the major record companies and major music publishers in the United States as of the time of Delivery of the Master concerned pursuant to Article 2 hereof but in no event later than the last date for timely Delivery of such Master pursuant to Article 2.

(B) On audio Phonograph Records distributed in Canada:

(I) If the copyright law of Canada provides for a minimum compulsory rate: The rate equal to seventy-five percent (75%) of the minimum compulsory license rate applicable to the use of musical works on audio Phonograph Records under the copyright law of Canada at the time of Delivery of the Master pursuant to Article 2 hereof but in no event later than the last date for timely Delivery of such Master pursuant to Article 2.

(II) If the copyright law of Canada does not provide for a minimum compulsory rate, but the major record companies and major music publishers in Canada (collectively the "Canadian Record Industry") have agreed to a mechanical license rate: The rate equal to seventy-five percent (75%) of the minimum license rate agreed to as of the time of Delivery of the Master concerned pursuant to Article 2 hereof but in no event later than the last date for timely Delivery of such Master pursuant to Article 2.

(III) If the copyright law of Canada does not provide for a minimum compulsory license rate, and the Canadian Record Industry has not agreed to a rate, the rate applicable under this clause (B) will be three and three-quarters cents ($.0375) (Canadian) per Composition.

(IV) The rate applicable under this clause (B) will not be more than the rate which would be applicable to the Records concerned under clause 10.01(a)(2)(A) above (Canadian) if they were manufactured for distribution in the United States.

(C) *On all other audio Records distributed in the United States*: The rate equal to seventy-five percent (75%) of the minimum compulsory license rate applicable to the use of musical compositions on such Records under the United States copyright law at the time of Delivery of the Master concerned but in no event later than the last date for timely Delivery of such Master pursuant to Article 2; provided, however, that (i) if, at the time any such Records are distributed, no such compulsory license rate has been implemented, then mechanical royalties for use of Controlled Compositions on such Records shall be paid after such compulsory license rate has been set under the United States copyright law, on all such Records (retroactively from the first such Record distributed hereunder); and (ii) if at any time legislation is enacted in the United States that expressly prohibits payment of less than one hundred percent (100%) of the minimum compulsory license rate, then solely with respect to the reproduction of Controlled Compositions on such Records, Company shall pay mechanical royalties at the minimum compulsory rate so prescribed by law for so long as such legislation remains in effect. The absence of any such compulsory license rate shall not impair the effectiveness of the license granted herein.

(D) *On all other audio Records distributed in Canada*: The rate equal to seventy-five percent (75%) of the minimum compulsory license rate applicable to the use of musical compositions on such under the copyright law of Canada (or the agreed to rate, as provided for in subparagraph 10.01(a)(1)(B)) at the time of Delivery of the Master pursuant to Article 2 hereof but in no event later than the last date for timely Delivery of such Master pursuant to Article 2; provided, however, that (i) if, at the time any such Records are distributed, no such compulsory license rate (and no agreed to rate, as provided for in subparagraph 10.01(a)(1)(B)) has been implemented, then mechanical royalties for use of Controlled Compositions on such Records shall be paid after such compulsory license rate has been set under the copyright law of Canada (or after an agreed to rate becomes generally applicable) on all such Records (retroactively from the first such Record distributed hereunder); and (ii) if at any time legislation is enacted in Canada that expressly prohibits payment of less than one hundred percent (100%) of the minimum compulsory license rate (or the agreed to rate), then solely with respect to the reproduction of Controlled Compositions on such Records, Company shall pay mechanical royalties at the minimum compulsory rate so prescribed by law for so long as such legislation remains in effect. The absence of any such compulsory license rate shall not impair the effectiveness of the license granted herein. The rights granted to Company herein include the rights to: (1) publicly perform any Controlled Composition by or through any means or manner not otherwise licensed by a performing rights society and (ii) incidentally reproduce or reproduce, in the form of server copies or other transient copies (solely to the extent such use is not otherwise licensed pursuant to a compulsory or voluntary license), any such Controlled Composition in connection with any transmission thereof. In addition, you hereby waive any so-called "moral rights" or any equivalent thereof otherwise available to you in connection with each such Controlled Composition.

(b) The total Mechanical Royalty for all Compositions (including Controlled Compositions) (i) embodied in or transmitted as a part of each Album other than Multiple Record Albums, will be not more than ten (10) times the Controlled Rate; (ii) embodied in or transmitted as a part of each single Record released hereunder, will be not more than two (2) times the Controlled Rate; (iii) embodied in or transmitted as a part of any EP released hereunder, will be not more than five (5) times the Controlled Rate; and (iv) embodied in or transmitted as a part of Multiple Record

Albums (if any), will be the maximum aggregate Mechanical Royalty will not be more than the maximum Mechanical Royalty applicable to an Album not in the form of a Multiple Record Album multiplied by a fraction, the numerator of which is the Suggested Retail List Price of such Multiple Record Album and the denominator of which is the Suggested Retail List Price of "top-line" Albums (deemed not to be less than fifteen dollars ($15.00)). With respect to the exploitation or sale of Records as described in paragraphs 6.02 (other than with respect to compact discs, Electronic Transmissions, EPs and Multiple Record Albums) and 6.03(a), the Controlled Rate and the Mechanical Royalty maximums will be three fourths (3/4) of the otherwise amounts prescribed above. Any amounts in excess of the applicable maximums pursuant to this subparagraph 10.01(b) or the applicable rates pursuant to subparagraph 10.01(a) above will be treated as described in subparagraph 10.01(g) below.

(c) Mechanical Royalties will not be payable for Controlled Compositions with respect to Records otherwise not royalty bearing hereunder, with respect to nonmusical material, with respect to Controlled Compositions of one minute or less in duration, and with respect to more than one (1) use of any one (1) Controlled Composition per Record. No Mechanical Royalties will be payable in respect of Compositions in the public domain or arrangements of Compositions in the public domain except that if such arrangement is credited by ASCAP, BMI, or SESAC then the Mechanical Royalty otherwise payable hereunder will be apportioned in the same ratio used by ASCAP, BMI, or SESAC in determining the credits for public performance of the work, provided you furnish Company with satisfactory evidence of that ratio.

(d) Company will compute Mechanical Royalties on Controlled Compositions as of the end of each calendar quarter-annual period in which there are sales or returns of Records on which Mechanical Royalties are payable to you. On or before the next May 15, August 15, November 15, or February 15, Company will send a statement covering those royalties and will pay any net royalties then due. If Company makes any overpayment of Mechanical Royalties on Controlled Compositions (e.g., but without limitation, by reason of an accounting error or by paying Mechanical Royalties on Records returned) such excess will be treated as described in subparagraph 10.01 (g) below. Reserves, the liquidation of reserves and your right to audit Company's books and records as the same relate to Mechanical Royalties for Controlled Compositions is subject to the terms and conditions set forth in Article 7.

(e) Any assignment made of the ownership of copyright in, or the rights to license or administer the use of, any Controlled Composition will be made subject to the provisions of this Article 10.

(f) With respect to Compositions (or portions thereof) which are not Controlled Compositions, you warrant and represent that Company and its designees shall be able to obtain mechanical licenses on rates and terms no less favorable than those contained in the standard mechanical license utilized by The Harry Fox Agency, Inc. in the United States and by CMRRA in Canada.

(g) You agree to indemnify and hold Company harmless from the payment of Mechanical Royalties in excess of the applicable amounts in the provisions of this Article 10. If Company pays any such excess, such payments will be a direct debt from you to Company, which, in addition to any other remedies available, Company may recover from royalties or any other payments hereunder.

11. *Failure of Performance*

11.01. Company will have the right to suspend the operation of this agreement and its obligations hereunder in the event Company is materially hampered in its recording, manufacture, distribution or sale of Records, or in the event its normal business operations become commercially impracticable, as the result of any cause beyond Company's control, including but not limited to labor disagreement, fire, earthquake, catastrophe, riot, shortage of materials, etc. If such contingency does not affect Company's ability to account to you and pay royalties then Company will account to you and pay royalties during any such suspension of this agreement. Such right may be exercised by written notice to you, and such suspension will last for the duration of the applicable event. A number of days equal to the total of all such days of suspension plus an additional seven (7) days will be added to the Contract Period in which such contingency occurs and the dates for the exercise by Company of its options as set forth in Article 1, the dates of commencement of subsequent Contract Periods, the date any other action is required hereunder, and the term of this agreement will be deemed extended accordingly. If such suspension of the term of this agreement exceeds six (6) consecutive months and affects no record manufacturer or distributor other than Company, you may, by notice to Company, request that Company terminate the suspension by notice given to you within thirty (30) days after its receipt of your notice. If Company does not do so, the term of this agreement will terminate at the end of such thirty (30) day period and all parties will be deemed to have fulfilled all of their obligations except those that survive the end of the term.

11.02. If Company wrongfully refuses to allow you to fulfill your Recording Commitment for any Contract Period, and if, not later than ninety (90) days after that refusal takes place, you notify Company of your desire to fulfill such Recording Commitment, then Company may permit you to fulfill such Recording Commitment by notice to you to such effect within thirty (30) days of Company's receipt of your notice. Should Company fail to give such notice, you will have the option to terminate the term of this agreement by notice given to Company within thirty (30) days after the expiration of the thirty (30) day period referred to above, and on receipt by Company of such notice the term of this agreement will terminate. If you fail to give Company either notice within the period specified in this paragraph 11.02, Company will be under no obligation to you for failing to permit you to fulfill such Recording Commitment. Alternatively, Company may notify you during any Contract Period that it does not intend to allow you to fulfill your Recording Commitment for the Period concerned, in which case the term of this agreement will terminate as of the date of such notice. In the event the term terminates under this paragraph, all parties will be deemed to have fulfilled all of their obligations hereunder except those obligations that survive the end of the term (e.g., warranties, re-recording restrictions and obligation to pay royalties), and Company will be obligated to promptly pay you, in full settlement of its obligations hereunder, an Advance in the amount equal to the Advance (or the balance thereof if a portion has already been paid to you) otherwise payable to you pursuant to paragraph 5.02 above with respect to the Album concerned.

12. *Representations, Warranties and Covenants*

12.01. You hereby represent, warrant and agree that during the term of this agreement, you will furnish to Company your exclusive recording services in the Territory.

12.02. You are authorized, empowered and able to enter into and fully perform your obligations under this agreement. Neither this agreement nor the fulfillment hereof by you infringes upon the rights of any Person. You have no knowledge of any claim or purported claim that may interfere with Company's rights hereunder or create any liability on the part of Company.

12.03. Intentionally Deleted.

12.04. As of the commencement of the term hereof, there are no unreleased recorded performances by you, other than those listed on Schedule A attached hereto, and incorporated by reference herein.

12.05. You warrant and represent that at Company's written request the Masters hereunder and performances embodied thereon will be produced in accordance with the rules and regulations of the American

Federation of Musicians, the American Federation of Television and Radio Artists and all other unions having jurisdiction. You warrant and represent that Artist is or promptly following Company's written request will become, and will remain to the extent necessary to enable the performance of this agreement, a member in good standing of all labor unions or guilds in which membership may be required for the performance of Artist's services hereunder.

12.06. You warrant and represent that you will not perform for any Person other than Company (and you will not license or consent to or permit the use by any Person other than Company of your name or likeness) for or in connection with the recording or exploitation of any Record embodying a Composition recorded by you under this agreement prior to the later of (i) the date five (5) years after the date of Delivery hereunder to Company of the last Master embodying that Composition, or (ii) the date two (2) years after the expiration or termination of the term of this agreement or any subsequent agreement between Company and you or any other Person furnishing your recording services. During the term or the period described in the first sentence of this paragraph 12.06, you will not perform or authorize the recording for use in advertisements of a Composition embodied on a Master Delivered hereunder without Company's prior written consent. Your agreement with the individual producer of each Master hereunder will restrict said producer from producing a Composition produced by such individual hereunder on another Master for any Person other than Company for at least two (2) years from the date of Delivery to Company of such Master.

12.07. You warrant and represent that none of the Masters hereunder, nor the performances embodied thereon, nor any other Materials, nor any authorized use thereof by Company or its grantees, licensees or assigns will violate or infringe upon the rights of any Person.

12.08. Without limiting the foregoing, Company will not be required to make any payments of any nature for or in connection with the acquisition, exercise or exploitation of rights pursuant to this agreement, except as specifically provided herein. Without limiting paragraph 3.04(c), you are solely responsible for and will pay all sums due the individual producers of each Master hereunder, and all other Persons entitled to receive royalties or other payments (excluding so-called "per record" union payments) in connection with the exploitation of Masters hereunder, including the sale of Records derived therefrom. You warrant and represent that each Person who renders any services in connection with the recording of Masters will grant to you and Company the rights referred to in this agreement and will have the right to so render such services and grant such rights. You warrant and represent that no Person whose performance is embodied in a Master hereunder or whose services are used in the recording of a Master hereunder (excluding those engaged by Company without your request that Company do so) will be bound by any agreement that may prevent or restrict such performances or services.

12.09. You will not authorize or knowingly permit your performances to be recorded for any purpose without an express written agreement prohibiting the use of such Recording on Records in violation of the restrictions herein. You will take reasonable measures to prevent the manufacture, distribution and sale at any time by any Person other than Company of such Records. Neither you, nor any Person deriving any rights from you, will use or authorize or permit any Person other than Company to (i) use any professional name you may adopt in connection with the exploitation of Masters recorded during the term hereof or, except as required under a prior agreement with an unrelated third party, in connection with the sale or other exploitation of Records during the term hereof; or (ii) except as required under a prior agreement with an unrelated third party, release or otherwise exploit Recordings recorded prior to the term hereof embodying your performances during the term hereof; or release or otherwise exploit Recordings embodying your performances recorded during the term hereof.

12.10. Neither you, nor any Person deriving any rights from you, will at any time do or authorize any Person to do anything inconsistent with, or that might diminish, impair or interfere with, any of Company's rights hereunder or the full and prompt performance of your obligations hereunder.

12.11. You are not under any disability, restriction or prohibition respecting Compositions recorded hereunder.

12.12. You hereby represent and warrant that you are above the legal age of majority pursuant to the laws governing this agreement and the performance hereunder.

13. *Company's Additional Remedies*

13.01. (a) Without limiting any other rights and remedies of Company hereunder, if you fail to Deliver any Masters hereunder within sixty (60) days after the time prescribed in Article 2, Company will have the following options, each exercisable by notice to you:

(i) Company may suspend its obligations to make payments to you under this agreement until you have cured the default; and/or

(ii) Company may terminate the term of this agreement at any time, whether or not you have commenced curing the default before such termination occurs; and

(iii) in the event Company terminates the term of this agreement under subparagraph 13.01(a)(2) above, you shall repay the amount not then recouped of any Advance previously paid by Company and not specifically attributable under paragraph 5.02 to an Album that has been Delivered, except as expressly provided in the next sentence. Notwithstanding the foregoing, you will not be required to repay any such amounts which you have spent on documented approved Recording Costs in connection with the Album that has not been delivered and which have been paid to third parties with whom you have no affiliation. For the avoidance of doubt, any such third party payments will be recoupable in accordance with paragraph 4.02 above.

(b) If Company terminates the term of this agreement under subparagraph 13.01(a)(2) above, all parties will be deemed to have fulfilled all of their obligations under this agreement except those obligations that survive the end of the term of this agreement [e.g., indemnification obligations, Company's obligation to account and pay royalties to you, re-recording restrictions, and your obligations under subparagraph 13.01(a)(3)]. No exercise of an option under this paragraph will limit Company's rights to recover damages by reason of your default, its rights to exercise any other option under this paragraph, or any of its other rights.

13.02. If your voice should be or become materially and permanently impaired, as determined by a physician selected by Company, or if you otherwise become physically unable to perform recording and/or personal appearances for a period in excess of six (6) consecutive months and/or if you cease to pursue a career as an entertainer, Company will have the right to terminate the term of this agreement by notice to you at any time during the period in which such contingency continues and thereby be relieved of any liability for the executory provisions of this agreement (other than Company's obligation to account and pay royalties when due).

13.03. You acknowledge, recognize and agree that your services hereunder are of a special, unique, unusual, extraordinary and intellectual character, giving them a peculiar value, the loss of which cannot be reasonably or adequately compensated for by damages in an action at law. Inasmuch as a breach of such services will cause Company irreparable damages, Company will be entitled to injunctive and other equitable relief, in addition to whatever legal remedies are available, to prevent or cure any such breach or threatened breach. Nothing in this agreement will prevent you from opposing such injunctive relief on any grounds that do not negate your acknowledgments in this paragraph.

13.04. The rights and remedies of Company as specified in this agreement are not to the exclusion of each other or of any other rights or remedies of Company. Company may decline to exercise one or more of its rights and remedies as Company may deem appropriate without jeopardizing any other of its rights or remedies. All of Company's rights and remedies will survive the expiration of the term of this agreement. Notwithstanding anything in this agreement, Company may at any time exercise any right it now has or at any time hereafter may be entitled to as a member of the public as though this agreement were not in existence.

13.05. (a) You agree to and do hereby indemnify, save and hold Company and its licensees harmless from any and all liability, loss, damage, cost and expense (including legal expenses and attorney fees) arising out of or connected with any breach or alleged breach of this agreement or any claim that is inconsistent with any of the warranties or representations made by you in this agreement. You agree to reimburse Company on demand for any payment made or incurred by Company with respect to the foregoing sentence, and, without limiting Company's rights or remedies, Company may deduct any amount not so reimbursed by you from any monies Company or an affiliate of Company owes you, whether hereunder or otherwise.

(b) Pending the determination of any claim in respect of which Company is entitled to be indemnified, Company may withhold monies otherwise payable to you hereunder in an amount not to exceed your reasonable potential liability to Company pursuant to this paragraph 13.05. At your written request, Company will release any such monies withheld if no legal action has been commenced on such claim, no settlement discussions have taken place and no further demand has been made on the claim for a period of at least one (1) year after the date of the last claim, demand or settlement discussion, whichever last occurred. If Company pays a claimant more than $ _____ _____ (the "Pre-authorized Amount") in settlement of any claim not reduced to judgment, you will not be obligated to reimburse Company for any of the settlement in excess of the Pre-authorized Amount unless you have consented to the settlement in writing. If you do not consent to a settlement proposed by Company for an amount exceeding the Pre-authorized Amount, you will nevertheless be required to reimburse Company for the full amount unless you make bonding arrangements, satisfactory to Company in its sole discretion, to assure Company of reimbursement for all damages, liabilities, costs and expenses (including legal expenses and counsel fees) that Company and its licensees may incur as a result of that claim.

(c) Company will promptly notify you of any action commenced on any claim subject to your indemnity hereunder. You may participate in the defense of any such claim through counsel of your selection at your own expense, but Company will have the right at all times, in its sole discretion, to retain or resume control of the defense of such claim.

14. *Definitions*

14.01. "Advance" – a prepayment of royalties. Advances are chargeable against and recoupable from any royalties (other than Mechanical Royalties except as otherwise set forth to the contrary herein) otherwise payable hereunder.

14.02. "Advertising and Promotion Purposes," when used in connection with Videos or Company Website Material – all uses for which Company receives no monetary consideration from licensees in excess of the costs of the Video or Company Website Material, an incidental fee, a reasonable amount as reimbursement for its administrative costs, and the actual costs incurred by Company in connection with such Videos or Company Website Material (e.g., for tape stock, duplication of the Videos and shipping).

14.03. "Album" or "LP" – a sufficient number of Masters embodying your performances to comprise one (1) or more compact disc Records, or the equivalent, of not less than forty (40) minutes of playing time and containing at least ten (10) different Compositions.

14.04. "Artist Domain Name" – a name embodying "_____" together with one or more other words mutually selected or approved by Company and you followed by ".XXX", as Internet Addresses. As used in the preceding sentence ".XXX" shall mean each and every so-called "second level" domain name now in existence or hereafter implemented including without limitation, ".com", ".net", ".org" together with territorial identifiers, e.g., ".UK".

14.05. "Ancillary Exploitations" – (a) the leasing of commercial advertising space to Persons other than Company or its licensees on the Company Artist Website; (b) the placement on the Company Artist Website of hyperlinks to so-called "ecommerce" Websites owned or controlled by Persons other than Company or its licensees; and (c) the inclusion of computer software, or Website links in ECD Material; and (d) Company's limited waiver of its exclusivity hereunder in granting a third party limited rights to broadcast a so-called "cybercast" of your performance via the Internet including any rebroadcasts of such performance, provided you and Company have approved such cybercast.

14.06. (a) "Artist Website" – Websites relating to you.

(b) "Company Artist Website" – the Artist Website created, maintained and/or hosted by Company or its licensees.

14.07. "Budget Record" – a Record which bears either: (a) a wholesale price equal to or less than two-thirds (2/3) of the highest wholesale price in the country concerned of top-line single-unit Records in the configuration concerned; or (b) a Suggested Retail List Price equal to or less than two-thirds (2/3) of the highest Suggested Retail List Price in the country concerned of top-line single-unit Records in the configuration concerned.

14.08. "Commercial Purposes" – when used in connection with Videos or Company Website Material – any use that is not for Advertising and Promotional Purposes (as defined above).

14.09. "Composition" – a single musical composition, irrespective of length, including all spoken words and bridging passages, including a medley.

14.10. "Contract Period" – the initial period, or any option period, of the term hereof (as such periods may be suspended or extended as provided herein).

14.11. "Container Charge" – ten percent (10%) of the Suggested Retail List Price for a single-fold analog disc Record in a standard sleeve with no insert; fifteen percent (15%) of the Suggested Retail List Price for an analog disc Record in a double-fold or gatefold jacket, in a nonstandard sleeve or jacket, or with inserts; twenty percent (20%) of the Suggested Retail List Price for analog cassette tape Records; and twenty-five percent (25%) of the Suggested Retail List Price for Records in the form of Compact Discs, Digital Compact Cassettes, Mini-Discs, Records sold in the form of other digital configurations, audiophile Records, Records sold in the form of any other new configurations, audiovisual Records, and for any other Record other than as herein provided; *provided, however*, there shall be no Container charge on Electronic Transmissions.

14.12. "Controlled Composition" – a Composition wholly or partly written, owned or controlled by you, a producer or any Person in which you or a producer has a direct or indirect interest.

14.13. "Delivery" – the receipt by Company of the fully mixed, edited and equalized Masters as provided in paragraphs 2.01 and 2.05 as well as the submission by you in written form of all necessary information, consents, clearances, licenses and permissions, including without limitation those relating to all samples, if any, interpolated in the Master Recordings, such that Company may manufacture, distribute and release the Records concerned, including, without limitation, all label copy, publishing and songwriting information (including, without limitation, applicable music performance rights organizations, and the names, addresses and telephone numbers of publishers), Album credits, the timings of and lyrics to each Composition contained on a Record, ancillary materials prepared by or for you which are required hereunder, first use mechanical licenses, if applicable, sideartist permissions, guest artist clearances, and any information required to be delivered to unions, guilds or other third parties.

14.14. "Developing Artist Series" – a program pursuant to which Company initially releases Through Normal Retail Channels in the United States, at a price (which may be effectuated by way of a rebate) equivalent to a Midline Record or a Budget Record, new Albums delivered to Company by one (1) or more of its developing artists.

14.15. "Digital Master" – a fully mixed, edited, equalized and leadered digital stereo tape master ready for the production of parts from which satisfactory Records can be manufactured.

14.16 "Distributor" – any Person authorized by Company to manufacture and distribute (or solely distribute) Company's Records.

14.17. "ECD Material" – all material acquired or created for inclusion in the "enhanced" or multimedia portion of an enhanced CD, CD Plus, CD ROM, DVD, or any other similar configuration, whether now known or hereafter created, (including, without limitation, Videos, photography, graphics, technology, software, so-called "hyperlinks" to Internet Addresses, etc.).

14.18. "Electronic Transmission" – any transmission to the consumer, whether sound alone, sound coupled with an image, or sound coupled with data, in any form, analog or digital, now known or later developed (including, but not limited to, "cybercasts," "webcasts," "streaming audio," "streaming audio/video," "digital downloads," "mobiletones," direct broadcast satellite, point-to-multipoint satellite, multipoint distribution service, point-to-point distribution service, cable system, telephone system, broadcast station, and any other forms of transmission now known or hereafter devised) whether or not such transmission is made on-demand or near on-demand, whether or not a direct or indirect charge is made to receive the transmission and whether or not such transmission results in a specifically identifiable reproduction by or for any transmission recipient. All references in this Agreement to the "distribution" of Records, unless expressly provided otherwise, shall be understood to include the distribution of records by way of Electronic Transmission thereof.

14.19. "Internet Addresses" – Uniform Resource Locators, addresses and/ or domain names.

14.20. "Joint Recordings" – Masters recorded pursuant to this agreement embodying your performance and any performance by another artist with respect to whom Company is obligated to pay royalties.

14.21. "Master," "Master Recording" or "Recording" – any recording of sound, whether or not coupled with a visual image, by any method and on any substance or material, whether now or hereafter known, that is, is intended to be, or could be embodied in or on a Record.

14.22. "Materials" – all Compositions embodied on a Master or Video hereunder; each name or sobriquet used by you, individually or as a group; and all other musical, dramatic, artistic and literary materials, ideas and other intellectual properties furnished or selected by you or any individual producer and contained in or used in connection with any Artist Website, Company Website Material, ECD Material, Recordings made hereunder or the packaging, sale, distribution, advertising, publicizing or other exploitation thereof.

14.23. "Mechanical Royalties" – royalties payable to any Person for the right to reproduce and distribute copyrighted musical compositions on Records.

14.24. "Midline Record" – a Record which bears either: (a) a wholesale price equal to more than two-thirds (2/3), but no more than eighty-five percent (85%), of the highest wholesale price in the country concerned of top-line single-unit Records in the configuration concerned; or (b) a Suggested Retail List Price equal to more than two-thirds (2/3), but no more than eighty-five percent (85%), of the highest Suggested Retail List Price in the country concerned of top-line single-unit Records in the configuration concerned.

14.25. "Mini Album" or "EP" – any Record, other than an Album, containing more than three (3) different Compositions.

14.26. "Multiple Record Album" – an Album containing two (2) or more cassettes, compact discs, or other configuration packaged as a single unit. For purposes of the Recording Commitment hereunder and for computing the applicable Authorized Budget or Advance, a Multiple Record Album accepted by Company will be deemed only one (1) Album.

14.27. "Net Receipts," "net sums," or "net amount received" and similar terms in this agreement – royalties or flat payments received by Company in connection with the subject matter thereof solely attributable to Masters or Videos hereunder, less all of Company's custom manufacturing, duplication, and packaging costs, less all advertising expenses and less any costs or expenses that Company incurs (such as, without limitation, production costs, Mechanical Royalties and other copyright payments, AF of M and other union or guild payments).

14.28. "Net Sales" – one-hundred percent (100%) of all sales of Records paid for and not returned (or such lesser percentage [not to be less than ninety percent (90%)] upon which Company's principal Distributor in the Territory concerned pays Company), less returns and credits, after deduction of reserves against anticipated returns and credits.

14.29. "Performance" – singing, speaking, conducting, or playing an instrument, alone or with others.

14.30. "Person" – any individual, corporation, partnership, association or other organized group of persons or legal successors or representatives of the foregoing.

14.31. "Phonograph Record" – a Record in a physical, non-interactive Record configuration (e.g., a vinyl disc, a cassette tape, a Compact Disc, a videocassette, etc.) as created by the manufacturer and/or distributor prior to its placement in distribution channels intended to reach the consumer.

14.32. "Recording Costs" – all amounts described in paragraph 4.01 above (other than so-called "per record" union payments) plus all other amounts representing direct expenses incurred by Company in connection with the recording of Masters hereunder (including, without limitation, travel, rehearsal, vocal coaching, musical instrument lessons, equipment rental and cartage expenses, costs incurred in connection with sampling, remixing and/or "sweetening", advances to individual producers, transportation costs, hotel and living expenses approved by Company, all studio and engineering charges, and all costs necessary to prepare Masters for release on all applicable media including those costs necessary to prepare final, equalized tapes therefor).

14.33. "Record" – all forms of reproduction, now or hereafter known, manufactured and/or distributed primarily for personal use, home use, school use, juke box use or use in means of transportation, including but not limited to sound-alone Recordings, audiovisual Recordings, interactive media (e.g., CD-ROM), and Electronic Transmissions.

14.34. "Royalty Base Price" – the Suggested Retail List Price less all excise, sales and similar taxes included in the price and less the applicable Container Charge.

14.35. "Sample(s)" or "sample(s)" – the embodiment of pre-existing Recording(s) and/or Composition(s) on a Master or Masters hereunder; provided, however, if all rights required for the purpose of manufacturing and distributing Records hereunder may be obtained by Company pursuant to a compulsory mechanical license such embodiment is not a Sample.

14.36. "Side" – a Recording of not less than three (3) minutes of continuous sound.

14.37. "Single" – a Record containing not more than three (3) different Compositions.

14.38. "Suggested Retail List Price" or "SRLP" –

(a) With respect to Records sold for distribution in the United States:

    (i) Other than with respect to Electronic Transmissions sold directly to consumer: Company's published suggested retail list price in the United States, it being understood that a separate calculation of the suggested retail list price will be made for each price configuration of Records manufactured and sold by Company.

    (ii) With respect to Electronic Transmissions sold directly to a consumer: the actual amount received by Company for such Records less any referral fees, commissions or similar fees payable to any Person unaffiliated with Company who, through their Website, electronic mail or other means, refers or directs to Company a purchaser of an Electronic Transmission or otherwise facilitates Company's sale to such consumer.

(b) With respect to Records sold for distribution outside the United States: the retail equivalent price utilized by Company's licensee in computing monies to be paid to Company for the Record concerned, provided that in any country where there is no actual suggested or applicable retail list price, the SRLP will be deemed to be the price established by Company or its licensee(s) as the retail equivalent price in conformity with the general practice of the recording industry in such country.

(c) Notwithstanding anything to the contrary contained herein, the Suggested Retail List Price for premium Records will be Company's actual sales price of such Records.

(d) Notwithstanding anything to the contrary herein, the Suggested Retail List Price with respect to so-called home video devices will be Company's published wholesale price for the device concerned.

(e) Notwithstanding anything to the contrary herein, the Suggested Retail List Price with respect to Records (other than Electronic Transmissions) sold by Company directly to a consumer through direct response, or otherwise will be Company's actual sales price of such Records.

14.39. The words "term of this agreement" or "period of this agreement" or "term hereof" or "so long as this agreement remains in force" or words of similar connotation refer to the initial period of this agreement and the period of all renewals, extensions and substitutions or replacements of this agreement.

14.40. "Territory" – the Universe.

14.41. "Through Normal Retail Channels" – Net Sales other than as described in paragraphs 6.02 (except that the fact that a Record is a compact disc will not in and of itself render such a sale not Through Normal Retail Channels provided it meets all other requirements therefor), 6.03, 6.04, 6.05, and 6.06. Notwithstanding the foregoing, Net Sales of a Multiple Record Album Delivered hereunder in fulfillment of your Recording Commitment with Company's written consent shall be deemed a sale Through Normal Retail Channels.

14.42. "United States" – the United States of America, its territories, possessions and military exchanges.

14.43. "USNRC Net Sales" – Net Sales Through Normal Retail Channels of the applicable Record sold for distribution in the United States.

14.44. "Video Costs" – any and all costs incurred by Company in connection with the production or exploitation of Videos and/or the acquisition of rights with respect thereto.

14.45. "Videos" – sight and sound Recordings that reproduce the audio performances of recording artists together with a visual image.

14.46. "Website" – a series of one (1) or more interconnected documents or files that are formatted using the Hypertext Markup Language, or any similar language, and that are intended to be accessible by Internet users.

14.47. (a) "Website Material" – all material acquired or created for inclusion on an Artist Website (including, without limitation, Videos, photography, graphics, technology, so-called "hyperlinks" to Internet Addresses, on-line chats, and electronic press kits or so-called "EPK"s).

   (b) "Company Website Material" – all Website Material for the Company Artist Website.

15. *Notices and Payments*

15.01. All notices required to be given to a party hereto must be in writing and sent to the address for the party first mentioned herein, or to such new address if changed as described below, in order to be effective. All royalties and royalty statements will be sent to you at your address first mentioned herein. Each party may change its respective address hereunder by notice in writing to the other. All notices sent under this agreement must be in writing and, except for royalty statements, may be sent only by personal delivery, registered or certified mail (return receipt requested), or by overnight air express (or courier shipment if outside the United States) if such service actually provides proof of mailing. The day of mailing of any such notice will be deemed the date of the giving thereof (except notices of change of address, the date of which will be the date of receipt by the receiving party). Facsimile transmissions will not constitute valid notices hereunder, whether or not actually received. A courtesy copy of any notice to Company shall be sent to _____

_____ .

16. *Miscellaneous*

16.01. Unless otherwise provided herein, as to all matters to be determined by mutual agreement and as to where any approval or consent by a party is required, such agreement, approval or consent may not be unreasonably withheld.

16.02. Unless otherwise provided herein, your agreement, approval or consent, whenever required, will be deemed to have been given unless you notify Company otherwise within five (5) business days following the date of Company's written request to you therefor.

16.03. The invalidity or unenforceability of any provision hereof will not affect the validity or enforceability of any other provision hereof. This agreement contains the entire understanding of the parties relating to its subject matter. No change of this agreement will be binding unless signed by the party to be charged. A waiver by either party of any provision of this agreement in any instance will not be deemed to waive it for the future. All remedies, rights, undertakings and obligations contained in this agreement are cumulative, and none of them are in limitation of any other remedy, right, undertaking or obligation of either party. Nothing contained herein will be construed so as to require the commission of any act contrary to law, and wherever there is any conflict between any provisions contained herein and any present or future statute, law, ordinance or regulation, the latter will prevail; but the provision of this agreement which is affected will be curtailed and limited only to the extent necessary to bring it within the requirements of the law.

16.04. Company has the right at any time during the term hereof to obtain insurance on your life, at Company's sole expense and cost, with Company being the sole beneficiary thereof. You agree that you will fully cooperate with Company in connection with the obtaining of such a policy, including, without limitation, submitting to any required physical examination and completing any documents necessary or desirable in respect thereof. Neither you nor your estate(s) have any right to claim the benefit of any such policy obtained by Company. If you fail your physical examination, such will not be a breach of this agreement, but thereafter Company will have the right to terminate the term hereof.

16.05. Company may assign its rights under this agreement in whole or in part only to any subsidiary, affiliated or controlling corporation, to any Person owning or acquiring a substantial portion of the stock or assets of Company, or to any partnership or other venture in which Company participates, and such rights may be assigned by any assignee. Company may also assign its rights to any of its licensees, if advisable in Company's sole discretion to implement the license granted. Neither you nor Artist shall be entitled to assign this agreement in whole or in part without the prior written consent of Company. Any assignment or purported assignment not authorized herein shall be null and void, and of no legal force or effect.

16.06. Neither you nor Company will be entitled to recover damages or to terminate the term of this agreement by reason of any breach by the other party of its material obligations hereunder unless the breaching party fails to remedy such breach within thirty (30) days following receipt of the non-breaching party's notice thereof. The foregoing cure period will not apply to your warranties hereunder, where a specific cure period is provided herein, to your obligation to Deliver Masters hereunder, to breaches incapable of being cured, or to an application for injunctive relief.

16.07. You recognize that the sale of Records is speculative and agree that the judgment of Company with respect to matters affecting the sale, distribution and exploitation of such Records is binding upon you. Nothing contained in this agreement obligates Company to make, sell, license or distribute Records manufactured from the Masters recorded hereunder except as specified herein.

16.08. This agreement has been entered into in the State of Tennessee. The validity, interpretation and legal effect of this agreement is governed by the laws of the State of Tennessee applicable to contracts entered into and performed entirely within such State. The Tennessee courts (state and federal), located in Nashville, will have exclusive jurisdiction over any controversies regarding this agreement, and the parties hereto consent to the jurisdiction of said courts. Any process in any action, suit or proceeding arising out of or relating to this agreement may, among other methods, be served upon you by delivering it or mailing it in accordance with Article 15 above. Any such process may, among other methods, be served upon Artist or any other Person who approves, ratifies, or assents in writing to this agreement to induce Company to enter into it, by delivering the process or mailing it to you or the other Person concerned in the manner prescribed in Article 15. Any such delivery or mail service will have the same force and effect as personal service.

16.09. In entering into this agreement and in providing services pursuant hereto, you have and will have the status of an independent contractor. Nothing herein contemplates or constitutes you as Company's agent or employee.

16.10. The headings of the Articles herein are intended for convenience only and will not be of any effect in construing the contents of this agreement.

16.11. This agreement will not become effective until executed by all parties hereto.

16.12. Any and all riders, exhibits or schedules annexed hereto together with this basic document constitute this agreement.

16.13. In the event of any action, suit or proceeding arising from or based on this agreement brought by either you or Company against the other, the prevailing party shall be entitled to recover from the other its reasonable attorneys' fees and costs in connection therewith in addition to any other relief to which the prevailing party may be entitled.

16.14. *You hereby acknowledge that you have been advised to seek the advice of independent legal counsel with regard to the interpretation and legal effect of this agreement and that you have had ample opportunity to do so and have either done so or have voluntarily relinquished your right to do so.*

17. *Leaving Member Provisions*

17.01. The term "you" as used in this agreement refers individually and collectively to the members of that group (whether presently or hereafter signatories to or otherwise bound by the terms of this agreement) currently professionally known as _____ (the "Group"). A breach of any term of this agreement or a disaffirmance or attempted disaffirmance of this Agreement on the ground of minority by or with respect to any member of the Group shall, at Company's election, be a breach by or with respect to the entire Group.

17.02. Individuals in addition to those presently members of the Group may become members of the Group only with Company's prior written approval. Additional members shall be bound by the terms of this Agreement relating to the you and you shall cause any additional member to execute and deliver to Company such documents as Company may deem necessary or desirable to evidence that individual's agreement to be so bound. You shall not, without Company's prior written consent, record any Master Recordings embodying the performances of an additional member prior to your delivery to Company of those documents, and if you do so, those Master Recordings, if Company so elects, shall not apply towards the fulfillment of your Recording Commitment.

17.03. If any individual member of the Group ("Leaving Member") ceases to be an actively performing member of the Group (e.g., that individual ceases regularly to record or perform live as a member of the Group or ceases regularly to. engage in other professional activities of the Group), you shall promptly give Company written notice thereof by certified or registered mail, return receipt requested. You shall designate a replacement member for that Leaving Member and Company shall have the right to approve of that replacement member. The replacement member shall be bound by the terms of this contract relating to you and you shall cause a replacement member to execute and to deliver to Company such documents as Company may deem necessary or desirable to evidence that replacement member's agreement to be so bound. You shall not, without Company's prior written consent, record any Master Recordings embodying the performances of a replacement member prior to your delivery to company of those documents, and if you do so, those Master Recordings, if Company so elects, shall not apply towards the fulfillment of your Recording Commitment.

17.04. (a) Company shall have the irrevocable option for the exclusive recording services of any Leaving Member. Company may exercise such option at any time by written notice to the Leaving Member but such notice shall be given no later than one hundred eighty (180) days after the date upon which Company shall have received the written notice required to be sent by you and referred to in paragraph 17.03 above. If Company shall so exercise its option with respect to any Leaving Member, that Leaving Member shall render his or her exclusive recording services to Company on the same terms and conditions contained in this agreement, except as otherwise hereinafter provided (the "Leaving Member Contract"):

   (i) The term of a Leaving Member Contract shall consist of an initial Contract Period commencing as of the date of Company's written notice to that Leaving Member pursuant to the preceding provisions of this paragraph 17.04 and shall continue until the <u>later</u> of (a) the last day of the twelfth (12th) month following the month of your Delivery to Company or (b) the last day of the ninth (9th) month following the month of Company's United States initial retail street date, of the Album Delivered by you in fulfillment of your Recording Commitment for the initial Contract Period, or such fewer number of days of which Company may notify the Leaving Member in writing. Company shall have the same number of options, each to extend the Term of the Leaving Member Contract for a Option Period, as equal the number of separate renewal options remaining under this agreement pursuant to Article 1 above as of the date that individual became a Leaving Member, but in no event shall Company have fewer than four (4) renewal options.

   (ii) Each Option Period under the Leaving Member Contract shall run consecutively and shall commence upon the expiration of the immediately preceding Contract Period thereunder and shall continue until the <u>later</u> of (a) the last day of the twelfth (12th) month following the month of your Delivery to

Company or (b) the last day of the ninth (9th) month following the month of Company's United States initial retail street date, of the Album Delivered by you in fulfillment of your Recording Commitment for the Contract Period, or such fewer number of days of which Company may notify the Leaving Member in writing. Company may exercise each option by giving the Leaving Member written notice of Company's election to do so at any time prior to the commencement of the Option Period for which it is exercised.

(iii) During the initial Contract Period and each Option Period of the Leaving Member Contract, the Leaving Member shall (in addition to any Demo Recordings required pursuant to paragraph 17.04(c) below) record and deliver to Company, at such times as Company shall designate or approve sufficient, Master Recordings embodying the Leaving Member's performances to constitute one (1) Album, plus, at Company's election, additional Master Recordings embodying that Leaving Member's performances, but in no event shall the Leaving Member be required to record for or deliver to Company in excess of twenty-four (24) Master Recordings during any particular Contract Period of the Leaving Member Contract.

(iv) The Advances set forth in Article 5 **[and the Authorized Budgets set forth in Article 3]** above shall not apply to the Master Recordings recorded by a Leaving Member under a Leaving Member Contract. Instead, Company shall pay the Recording Costs of the Master Recordings recorded at recording sessions conducted in accordance with the terms of the Leaving Member Contract in an amount not in excess of the recording budget therefor approved by Company in writing.

(v) With respect to Master Recordings embodying the performances of a Leaving Member recorded pursuant to a Leaving Member Contract, the royalty rates pursuant to paragraph 10.01 shall be **[seventy-five (75%) percent]** of the otherwise applicable royalty rates.

(vi) If your account is in an unrecouped position as of the date of Company's written notice to that Leaving Member, an amount equal to all unrecouped advances or charges against royalties pursuant to this agreement multiplied by a fraction, the numerator of which shall be one (1), and the denominator of which is the total number of individual members of the Group as of the date of your notice to Company pursuant to paragraph 17.03 hereof (the "Fraction), shall be deemed an advance with respect to a particular Leaving Member, recoupable from royalties payable by Company under the Leaving Member Contract with that Leaving Member; provided, however, that to the extent Company recoups any such portion of the unrecouped balance from the Leaving Member's royalties, the amounts so recouped shall be credited to your account hereunder. Further, Company shall have the right to recoup from the Leaving Member's share (if any) of royalties payable to you under this agreement any such advances or charges under the Leaving Member Contract; however, in no event shall Company have the right to recoup any unrecouped balance more than once.

(b) At Company's request, you shall cause any Leaving Member to execute and deliver to Company any and all documents as Company may deem necessary or desirable to evidence the foregoing, including, without limitation, an exclusive recording contract with Company relating to that Leaving Member's recording services.

(c) If Company shall enter into a Leaving Member Contract with a particular Leaving Member, that Leaving Member shall, upon Company's request, record and deliver to Company at such times as Company shall indicate no fewer than four (4) so-called "Demo Recordings," each embodying that Leaving Member's performance as the sole featured artist of a single Composition previously unrecorded by that Leaving Member and approved by Company and each Demo Recording shall be recorded in its entirety in a recording studio. Company shall pay the costs of the Demo Recordings at recording sessions conducted in accordance with the terms hereof in an amount not in excess of a recording budget therefor approved by Company in writing. All Recording Costs paid by Company in connection with the recording of the Demo Recordings shall be recoupable by Company from royalties payable by Company under the Leaving Member Contract or under any other agreement between you and Company or Company's affiliates. Notwithstanding anything to the contrary contained herein, Company may, at Company's election, terminate Company's Leaving Member Contract with a Leaving Member, upon sending that Leaving Member written notice of Company's election to do so within sixty (60) days after Company's receipt of the completed Demo Recordings, and thereby be relieved of any obligations or liabilities under that Leaving Member Contract. Company shall own the Demo Recordings and all reproductions and derivatives thereof to the same extent that Company own the Masters.

17.05. Notwithstanding any of the foregoing, if any member of the Group shall be a Leaving Member or if the Group shall completely disband, Company may, without limiting Company's other rights and remedies, terminate the Term of this agreement by written notice to you and shall thereby be relieved of any obligations or liabilities hereunder, except Company's obligations with respect to Masters recorded prior to that termination. In the event Company elects to so terminate the Term of this agreement, paragraph 17.03 above shall be applicable to each member of the Group as if each member were a Leaving Member.

17.06. **[If a Key Member (as defined below) shall become a Leaving Member and Company does not terminate the Term of this agreement, then with respect to each Album delivered under the agreement subsequent to the date on which that Key Member became a Leaving Member, the Advances set forth in Article 5 and the royalty rates set forth in Article 6 above shall be reduced by multiplying the advance amounts and the royalty rates, respectively, by a fraction, the numerator of which is the number of individuals in the Group after the Leaving Member became a Leaving Member, and the denominator of which is the number of individuals in the Group prior to the Leaving Member becoming a Leaving Member. As used herein, the term "Key Member" shall mean a member of the Group whose contribution to the Group, in the recording studio, at live performances, as a songwriter or otherwise, is, in Company's opinion, material.]**

17.07. If any member of the Group shall become a Leaving Member, that member shall not have the right thereafter during the Term to use any name utilized by the Group or any name similar thereto. Without limiting the generality of the foregoing, that member shall not, in connection with any of his or her professional activities, use the phrase (formerly a member of ("any name used by the group")) or any similar expression.

17.08. All notices, statements or other correspondence to a Leaving Member shall be sent by Company to your address above, or at such other address of which that Leaving Member shall have advised Company in writing.

17.09. Unless Company receives instruction from you in writing otherwise, Company shall pay all royalties and advances due you hereunder to: _____ c/o _____ .

_____

By: _____   _____

   An Authorized Signatory

Social Security #: _____

SCHEDULE A

Ke$ha's artist management contract

The following is a transcript of Ke$ha's original management contract with DAS Communications.

DAS Communications, Ltd. Date: 1/27/2006
83 Riverside Drive
New York, NY 10024
Gentlemen:

I have carefully considered the advisability of obtaining your assistance and guidance in the furtherance of my musical, recording, artistic, theatrical and literary career, and have made independent inquiry concerning your ability and reputation. I have determined that your services would be of great value in the furtherance of my career because of your extensive knowledge of and reputation in the entertainment and amusement industries. I have, therefore, determined that I wish you to act as my exclusive personal manager for the term of this agreement, subject to the following provisions and conditions:

1. I hereby engage you as my exclusive personal manager throughout the world commencing as of the date hereof for a term of five (5) years, and you hereby accept such engagement.

2. You agree to use your best efforts to perform the following services for me:

   a. To represent me and act as my advisor in all business negotiations and matters of policy relating to my career; to consult with me regarding my engagements and to consult with employers to assure, to the best of your ability, the proper use of my services; to advise and counsel me in the selection of literary, artistic and musical material, in matters relating to publicity, public relations and advertising, in the adoption of the proper format for presentation of my talents and to arrange for interviews or auditions designed to further my career.

   b. To cooperate with and consult with me regarding my relations with any theatrical and literary agents employed, to advise and counsel me in the selection of third parties to assist, accompany or improve my artistic presentation and to make yourself available at reasonable times at your office to confer with me in all manners concerning my career.

3. You are authorized and empowered for me and on my behalf after obtaining my prior approval to do the following: approve and permit any and all publicity and advertising; approve and permit the use of my name, photograph,

likeness, voice, sound effects, caricatures, literary, artistic and musical materials for the purposes of advertising and publicity and in the promotion and advertising of any and all products and services; negotiate for me on my behalf any and all agreements, documents and contracts for my services, talents and/or artistic, literary and musical materials. You are not required to make any loans or advances to me or for my account, but, in the event you do so, I shall repay them promptly. The authority herein granted to you is coupled with an interest and shall be irrevocable.

4. I shall at all times engage and utilize proper theatrical agents or employment agencies to obtain engagements and employment for me. I shall advise you of all offers of employment submitted to me and will refer any inquiries concerning my services to you, in order that you may advise me whether the same are compatible with my career. (IT IS CLEARLY UNDERSTOOD THAT YOU ARE NOT AN EMPLOYMENT AGENCY OR A THEATRICAL AGENT, THAT YOU HAVE NOT OFFERED OR ATTEMPTED OR PROMISED TO OBTAIN, SEEK OR PROCURE EMPLOYMENT OR ENGAGEMENTS FOR ME, AND THAT YOU ARE NOT OBLIGATED, AUTHORIZED, LICENSED OR EXPECTED TO DO SO.)

5. This agreement shall not be construed to create a partnership between us. It is specifically understood that you are acting hereunder as an independent contractor and you may appoint or engage any and all other persons, firms or corporations throughout the world in your discretion to perform any or all services which you have agreed to perform hereunder. Your services hereunder are not exclusive and you shall at all times be free to perform the same or similar services for others as well as engage in any and all other business activities so long as the same does not materially interfere with the services to be rendered by you hereunder. You shall only be required to render reasonable services which are called for by this agreement as and when reasonably requested by me. You shall not be required to travel or to meet with me at any particular place or places except upon my request and then only in your discretion and at my expense.

6. In consideration of your agreement hereto and as compensation for services rendered and to be rendered to me by you hereunder, I agree to pay you as and when received by me a sum equal to twenty (20%) percent of any and all gross monies or other considerations which I may receive as a result of my activities in and throughout the entertainment, amusement, music and recording industries, including any and all sums resulting from the use of my artistic and literary talents and the results and proceeds thereof. Without in any manner limiting the foregoing, the matters upon which your compensation shall be computed shall include any and all of my activities in connection with matters as follows: motion pictures, television, radio, music, literary, theatrical engagements, personal appearances, public appearances in places of amusement and entertainment, records and recordings, publications, and the use of my name, approved likeness and talents for purposes of advertising

and trade. Subject to Paragraph "16" hereof, I likewise agree to pay you a similar sum following the expiration of the term hereof upon and with respect to any and all engagements, contracts and agreements entered into or substantially negotiated for during the term hereof relating to any of the foregoing, and upon any and all extensions, renewals and substitutions thereof and upon any resumptions of such engagements, contracts and agreements entered into within six (6) months after the end of the term. Subject to Paragraph "15" hereof, the term "gross monies or other considerations" shall include, without limitation, salaries, earnings, fees, royalties, gifts, bonuses, shares of profit, shares of stock, partnership interests, percentages and the artist's share of the total amount paid for a package television or radio program (live or recorded), motion picture or other entertainment packages, less any production expenses incurred by me, earned or received directly or indirectly by me or my heirs, executors, administrators or assigns, or by any person, firm, corporation on my behalf. In the event that I receive, as all or part of my compensation for activities hereunder, stock or the right to buy stock in any corporation or in the event that I receive property, whether as individual proprietor, stockholder, partner, joint venture or otherwise or in the event that I shall cause a corporation to be formed, your percentage shall apply to my said stock, individual proprietorship, partnership, joint venture or other form of interest, and you shall be entitled to your percentage share thereof. Should I be required to make any payment for such interest, you will pay your percentage share of such payment, unless you decline to accept your percentage share thereof.

7. Notwithstanding anything to the contrary contained herein, I shall have the right to terminate this agreement if a recording agreement providing for my services as a recording artist, to be distributed by a major label, (the "Recording Agreement") has not been entered into within one (1) year from the date hereof. In the event I do enter into such an agreement, the term of this agreement shall extend until the later of (a) five (5) years from the date hereof or (b) ending upon the completion of the Second Album Cycle under said recording agreement. For purposes hereof, the term "Second Album Cycle" shall mean the period beginning upon the commencement of pre-production of the Second Album and ending upon the completion of all touring and personal appearances in support of the Second Album. In the event I do enter into a recording agreement and it terminates prior to the end of the term hereof, and a period of six (6) months elapses without my entering into a new recording agreement with a major label, then I shall have the right to terminate this agreement.

8. Upon written notice by either of us to the other, the party to whom such notice is addressed will furnish an accounting to the other party of all transactions between us and all transactions by you on my behalf within thirty (30) days of such request.

9. In addition to the sums required to be paid to you as aforesaid, I shall reimburse you exclusively on my behalf or in connection with my career or in the performance of your services hereunder which are substantiated by receipted vouchers or paid bills. You shall not incur any single expenditure in excess of $500 or monthly expenditures in excess of $1,000 without my prior approval. In this regard, you agree to send me monthly statements of account.

10. In order to make specific and definite and/or to eliminate, if possible, any controversy which may arise between us hereunder, we each agree that if at any time either of us feels that the terms of this agreement are not being performed as herein provided, we each will so advise the other in writing by Registered or Certified Mail, Return Receipt Requested, of the specific nature of the claim, non-performance or mal-performance and shall allow a period of thirty (30) days after receipt thereof within which to cure such claimed breach. No such legal proceedings may commence prior to the expiration of the aforesaid thirty (30) day period. In order to be effective, a copy of any notices to me must be sent simultaneously to Goldring, Hertz & Lichtenstein, LLP, 450 N. Roxbury Drive, 8th floor, Beverly Hills, CA 90212, Attn: Fred Goldring, Esq.

11. I warrant and represent that I am over eighteen (18) years of age, and that I am free to enter and perform under this agreement.

12. We each agree to indemnify and hold the other safe and harmless from any and all loss or damage including reasonable out-of-pocket attorneys' fees arising out of or in connection with any claim by a third party which is consistent with any of the warranties, representations, covenants or agreements mad by the indemnifying party herein. Our joint indemnifications herein shall be limited to actual final non-appealable judgments and/or claims which are settled with both parties' consent, which we agree not to unreasonably withhold.

13. I agree to pay or cause to be paid any and all monies due you to you not later than thirty (30) days after my receipt of any such monies. In this regard, I agree that you shall have the right, upon reasonable written notice to me, to review my books and records as same pertain to the subject matter hereof provided that you shall not have such right to audit my books and records more than once in a calendar year.

14. Notwithstanding anything to the contrary contained herein, you agree that you will not execute on my behalf any agreements calling for my services in the entertainment industry. All such agreements shall be executed by me although I agree that when I am unavailable to do so, you shall have the right to execute on my behalf subject to your consulting with my attorney, and subject to my approval of the terms, all modifications of the recording agreement but only as same pertain to marketing and promotion of any of my recordings and you deem such modification beneficial to my career. You shall of course have the right, on my behalf, to enter into day to day booking contracts for so-called "one-nighters" or a short series of "one-nighters."

15. Notwithstanding anything to the contrary contained herein you agree not to commission recording costs, recoupable video expenses, recoupable video expenses, recoupable tour support payments from my record company up to the amount required to deficit finance a tour, third party production costs, the costs of any opening acts, passive investments, per diem payments to me, gifts to me (except those given in lieu of monetary payments) and non-entertainment activities. You also agree not to commission sound and light production reimbursements with live personal appearance contracts.

16. Notwithstanding anything to the contrary contained herein, with respect to post term earnings from any agreement which I may enter into during the term of this agreement, your commission shall be ten (10%) percent for the two (2) year period following the end of the term hereof and zero (0%) percent thereafter, provided, however, that you shall receive full commissions pursuant to Paragraph "6" in perpetuity on all recordings which are commercially released during the term and with respect to all songs which are published during the term. Songs shall be deemed to have been published if a recording of the song is commercially released during the term or if ant synchronization income has been generated by way of a license of the song during the term. With respect to recordings made during the term and commercially released within two (2) years after the term and/or songs written during the term and published within two (2) years after the term, you shall receive a commission of ten (10%) percent in perpetuity. You shall receive no commission on recordings made and released after the term and/or songs written and published after the term.

17. Notwithstanding anything to the contrary contained herein, provided that I appoint a reputable business manager experienced in entertainment industry accounting practices (a "Business Manager") who shall be instructed to remit monies due you hereunder, you agree that monies due me may be paid directly to said business manager rather than as is otherwise provided for herein.

18. Notwithstanding anything to the contrary contained herein, I agree to direct my Business Manager to make payment of all monies due you hereunder directly to you. I hereby further agree that with respect to post term commissions which are due you I shall instruct any and all persons, firm or corporations who may be holding monies otherwise due you to irrevocably pay same directly to you. I agree to execute letters of direction to effectuate said irrevocable payments.

19. This agreement is made and executed in the State of New York and shall be construed in accordance with the laws of said State applicable to contracts wholly to be performed therein. In the event that any provision hereof shall, for any reason, be invalid or unenforceable, the same shall not affect the validity or enforceability of the remaining provisions hereof. This agreement is the only agreement between us concerning the subject matter hereto and may not be modified or terminated except by an instrument in writing executed by the parties hereto. A waiver by either of us of a breach of

provision hereof shall not be deemed a waiver of any subsequent breach, whether of a similar or dissimilar nature.

If the foregoing correctly states the terms of our understanding, will you kindly so indicate by signing below at the place indicated.

Very truly yours,

(Signed)

Kesha Rose Sebert

ACCEPTED AND AGREED TO:

DAS Communications, Ltd.

By: ___(Signed)_____

An authorized signatory

This is a transcript of the original artist management agreement filed as part of a lawsuit with the Supreme Court in New York, accessed at this web address: https://iapps.courts.state.ny.us/fbem/DocumentDisplayServlet?documentId=Wi6A TK7yODWVmYDO0ytqOw==&system=prod

Code of conduct: Music manager's forum in Australia

As we noted elsewhere in this book, there is no association of artist managers in the United States and therefore no organization that provides professional conduct advice to those who perform management services to artists in the music business. The Music Manager's Forum in Australia has adopted the following guidelines into its constitution, saying that managers must at all times and to the best of their ability conduct themselves with the following principles:

1. Devote sufficient time so as to properly fulfill the requirements of good management in the interest of the artists;
2. Not act in any fashion, which is detrimental to their clients' interests;
3. Conduct themselves in a manner which is professional and ethical, and which abides by best Business practices and methods accepted in their country and comply with any Statutory Regime that is in place or is created.
4. Conduct all of their affairs with their clients in a transparent manner;
5. Protect and promote the interest of their clients to the highest possible standard;
6. Exercise the rights and powers implied or granted to them by their clients in the written or oral agreements for the clients' best interest.
7. Ensure that no conflict of interest shall interfere with the discharge of their duties towards their clients.
8. All conflicts of interest must be disclosed immediately and noted in any artist agreement.
9. The Manager's share of the proceeds coming from the artist's professional (artistic) activity may not exceed 25% of the artist's income.
10. The Manager must ensure that all monies (income and expenditure) due to the artist are managed completely separately to the private assets of the manager.
11. The Manager makes a commitment (and is duty bound) to absolute transparency in all contractual and financial business dealings that concern collaboration with the artist. This includes in particular giving access to all accounting, settlement of accounts with third parties, and contracts.
12. Should a member be proved to have breached the Code of Conduct, they may be debarred (expelled) from the Organization. The expulsion process is decided according to the Constitution. In the event of an expulsion from the

organization, the particular member is no longer entitled to use MMF membership credentials and the Organization is entitled to advise the membership, the IMMF; its affiliates or associated entities, including Government and Industry bodies.

13. Should the individual seek or renew such membership of any other organization affiliated with the IMMF, MMF will use its utmost powers to ensure such a member will not be granted membership, or if membership granted it is conditional on that member's behavior and adherence to the Code.

14. Music Managers shall respect the integrity of other managers in their relationships with their Artists and not actively interfere with same.

15. A manager who is approached by an artist, who was previously the client of another manager, shall confirm that the artist has fulfilled his, her, or their legal obligations to the previous manager before entering into a management relationship with the artist.

16. Where a manager acts as publisher, agent, record producer or in any other capacity as well as a Manager for his, her, or their clients, they shall declare such interests so that the artist has the ability to determine for themselves if they feel it constitutes a conflict of interest and therefore detrimental to the artist's career.

17. Where a manager acts in any other capacity as well as manager for his, her or their clients where such activity ordinarily involves the charging of fees or commissions, the manager shall not charge multiple fees or commissions, instead charging either the agreed management commission alone or the fee or commission usually charged for that other activity and forgoing their management commission. Where the manager elects to charge a fee or commission other than the management commission they shall first gain the consent of their artist in writing.

18. Managers must ensure that all monetary transactions made on behalf of or in the interest of the client and all books of account and records must always be reasonably open for the inspection of the artist or their appointed representative with reasonable notice, during business hours.

19. Where a manager engages an artist under a written agreement, the manager shall ensure that their client seeks and receives expert legal advice on the terms of such agreement before signing it.

20. Managers will endeavour to keep themselves well informed of current events and legislation, both national and international, as it pertains to the proper exploitation of their client's career and the proper administration of their client's business/s.

Used by permission of the Music Managers Forum of Australia, with special thanks to MMF Chairman Nathan Brenner.

Index

Page numbers followed by *f* indicate figures and *t* indicate tables.